POLYMER THIN FILMS

POLYMER THIN FILMS

Editor

Eden Slaver

scitus
academics

Polymer Thin Films
Edited by **Eden Slaver**

ISBN: 978-1-68117-601-7
Library of Congress Control Number: 2016946456

SCITUS Academics LLC,
616, Corporate Way, Suite 2, 4766,
Valley Cottage, NY 10989

www.scitusacademics.com

Notice

Printed in United States of America on Acid Free Paper

PREFACE

Thin polymer films have many technological applications in different industrial and biomedical sectors related to functional and protective coatings, non-fouling bio-surfaces, medical implant biocompatibility, separation, advanced membranes, microfluidics, devices and sensors, adhesion, lubrication and friction alteration. Polymer thin film is an emerging area driven by its enormous technological potential and its academic problems.

This book covers the current topics of polymer films produced by leading industry experts. Contributors come from a wide range of disciplines, engineers, material scientists, chemical engineers, including chemists and physicists, to reflect the interdisciplinary nature of the field. This book gives an up-to-date overview of the current knowledge of the use of polymer thin film for important technological use. Polymer Thin Films covers the scientific principles and technologies required for the application of electronic polymer devices. Both academic and practical scientists have a broad and definitive coverage of this emerging field. The book is designed to allow readers with a particular background, for example, nanotechnology polymer, to become familiar with other specialist aspects of this multidisciplinary field.

TABLE OF CONTENTS

Chapter 1 Preparation and Characterization of Silica/Polyamide-imide Nanocomposite Thin Films 1

Chapter 2 Temperature- and thickness-dependent elastic moduli of polymer thin films 17

Chapter 3 Improving the Performances of Random Copolymer Based Organic Solar Cells by Adjusting the Film Features of Active Layers Using Mixed Solvents 31

Chapter 4 Synthesis and characteristics of NH_2-functionalized polymer films to align and immobilize DNA molecules 55

Chapter 5 Electrospray deposition of polymer thin films for organic light-emitting diodes 71

Chapter 6 Characterization of Electrochromic Properties of Polyaniline Thin Films Electropolymerized in H_2SO_4 Solution 85

Chapter 7 Preparation and Characterization of Subsurface Silver Particulate Films on Polymer Blends of Polystyrene/Poly (2-vinylpyridine)/ Poly(vinylpyrollidone)/ Poly(4-vinylpyridine) 101

Chapter 8 Functionalized Cellulose Networks for Efficient Oil Removal from Oil–Water Emulsions 121

Chapter 9 Cohesive strength of nanocrystalline ZnO:Ga thin films deposited at room temperature 145

Chapter 10 A study on the characteristics of plasma polymer thin film with controlled nitrogen flow rate 157

Chapter 11 Magnetoelectricity in $CoFe_2O_4$ nanocrystal-P(VDF-HFP) thin films 169

Chapter 12 Polymerization of Thin Film Polymers 195

Chapter 13 Complex microstructures of ABC triblock copolymer thin films directed by polymer brushes based on self-consistent field theory 231

Chapter 14 Photothermal Effects and Applications of Polydimethylsiloxane Membranes with Carbon Nanoparticles 259

Index 293

7

PREPARATION AND CHARACTERIZATION OF SILICA/POLYAMIDE-IMIDE NANOCOMPOSITE THIN FILMS

Xiaokun Ma, Nam-Hee Lee, Hyo-Jin Oh, Jong-Sun Hwang and Sun-Jae Kim

ABSTRACT

The functional silica/polyamide-imide composite films were prepared via simple ultrasonic blending, after the silica nanoparticles were modified by cationic surfactant—cetyltrimethyl ammonium bromide (CTAB). The composite films were characterized by scanning electron microscope (SEM), thermo gravimetric analysis (TGA) and thermomechanical analysis (TMA). CTAB-modified silica nanoparticles were well dispersed in the polyamide-imide matrix, and the amount of silica nanoparticles to PAI was investigated to be from 2 to 10 wt%. Especially, the coefficients of thermal expansion (CET) continuously decreased with the amount of silica particles increasing. The high thermal stability and low coefficient of thermal expansion showed that the nanocomposite films can be widely used in the enamel wire industry.

KEYWORDS

Composites, Fracture, Surfaces, Thermal properties, Thin films

1. INTRODUCTION

Polyamide-imide (PAI) is a kind of thermoplastic resin, has good high-temperature resistance, outstanding mechanical properties and excellent oxidative stability, all of which have led PAI to have been widely used with electronic materials, adhesives, composite materials, fiber, and film material [1–8]. Comparing with the polyimide and polyamide, the PAI own the better process ability and heat-resistant properties. The basic studies of the synthesized PAI have attracted increasing interest for a wire-coating material with a thermal resistant property. However, an increasing number of breakdown cases have been reported when a higher surge voltage exists in some devices. Therefore, the development of an inorganic/organic nanocomposite insulating material should be widely discussed to improve the electrical life of enameled wires [9–11].

Actually, there are so many factors affecting the properties of hybrid composites, such as the particle size, size distribution, and filler content. In addition, the inorganic particle shape, surface structure, and mechanical properties of a filler (stiffness, strength, etc.) all play an important role in the synthesizing of inorganic/organic composite materials. Especially, the bond strength between inorganic particles and polymer matrix should be improved, which always was influenced by the type of dispersion aid or coupling agent used [12–18].

Silica nanoparticles as a very important inorganic material have emerged as an area of intense current interest motivated because of their special physical and chemical properties, such as their small size, strong surface energy, high scattered performance, and thermal resistance [19–23]. However, the applications of silica nanoparticles are largely limited because of their high energetic hydrophilic surface, which causes the silica nanoparticles to be easily agglomerated. Fortunately, this problem could be resolved by using some surface modification methods with different surfactant agents. In other words, the strong interface adhesion between the organic matrix and silica nanoparticles is a key to the application of silica nanoparticles as fillers.

Jadav et al. [24] had successfully synthesized the silica/polyamide nanocomposite film via interfacial polymerization process using two types of silica nanoparticles of size about 16 and 3 nm, respectively. The nanocomposite films exhibit superior thermal stability than the pure polyamide membranes. In this work, the authors observed that silica nanoparticles loading could significantly modify the polyamide network structure, pore structure and transport properties. The excellent membrane performance in terms of separation efficiency and productivity flux was discussed. Zhang et al. [25] prepared a novel isometric polyimide/silica hydrid material through sol–gel technique. At the beginning, 3-[(4-phenylethynyl) phthalimide]propyl triethoxysilane (PEIPTES) was synthesized to be used for modifying nano-silica precursor. Then the isomeric polyimide/silica hybrid material was produced by using isomeric polyimide resin solution and the modified nano-silica precursor after heat treatment process. The isomeric polyimide/silica composite has much better thermal properties and nano-indenter properties than those of isomeric polyimide.

In this work, cationic surfactant CTAB was chosen to modify the silica nanoparticles and the amount of CTAB to silica nanoparticles was changed from 0 to 3 wt%. After the surface modification process, the CTAB-modified silica nanoparticles had the better compatibility with PAI polymer matrix and the CTAB-modified silica nanoparticles could be well dispersed into the PAI matrix, even when the amount of silica to PAI reached to 10 wt%. The thermal properties were obviously improved, and the decomposition temperature was increased with the amount of silica increasing.

2. EXPERIMENT

2.1 Materials

Silica nanoparticles were purchased from Sigma–Aldrich Company, and the size of the nano-silica was 10–20 nm. Commercial polyamide-imide powder (PAI, Torlon AI-10) was purchased from Solvay Company (USA). Cetyltrimethyl Ammonium Bromide (C16, CTAB, 99% purity, ACROS) was used as a surface modification agent. *N,N*-Dimethylformamide (DMF) was anhydrous and was purchased from

Sigma–Aldrich Company. Sodium hydroxide (NaOH) was purchased from Sam Chun Co (Korea). Ethanol was a chemical reagent and was purchased from DUKSAN Pure Chemicals. Distilled and deionized water was used throughout the work.

Bar coater 08, coating rods are used primarily to apply a variety of coatings or emulsions to a multitude of substrates. Stainless steel rods are wrapped very tightly with stainless steel wire and the diameter of wire are 0.2 mm. In this work, the silica/PAI composite films were cast by bar coater to make the films more uniform.

2.2 Preparation and Surface Modification of Silica Nanoparticles

In a typical experiment, 100 ml deionized water and 1.000 g silica particles were added to the flask, and the solution was adjusted to pH 8 by the addition of 0.1 M NaOH. Then the silica nanoparticles were modified at 65°C under constant stirring with CTAB added into the system. In order to improve the dispersal state of the silica nanoparticles in the PAI matrix, the amount of CTAB to silica nanoparticles was changed from 0 to 3.0 wt%. After the surface modification process, the modified nano-silica particles were collected by suction filtration and then dried at 90°C for 6 h.

2.3 Preparation of Silica/Polyamide-imide Nanocomposite Films

The silica/PAI nanocomposite films were prepared through the simple ultrasonic blending. *N,N*-dimethyl form amide (DMF) was used as a solvent and 2.0 g of PAI powder were dissolved in 3 ml of DMF. The silica nanoparticles were added into the solution, the amount of silica to PAI was changed from 2 to 10 wt%. The mixture was ultrasonic dispersed ~3 h at room temperature. In the next step, the mixture solution was cast on a square glass plate (5*5 cm) by the bar coater. The films were initially heated to remove the solvent in the vacuum oven and the temperature was controlled as follows: at 60°C for 4 h, at 80°C for 2 h, then raised to 120°C for 1 h; finally, the films were heated at 160°C for 1 h. The experiment details of the process of silica/PAI composite films are shown in Scheme 1.

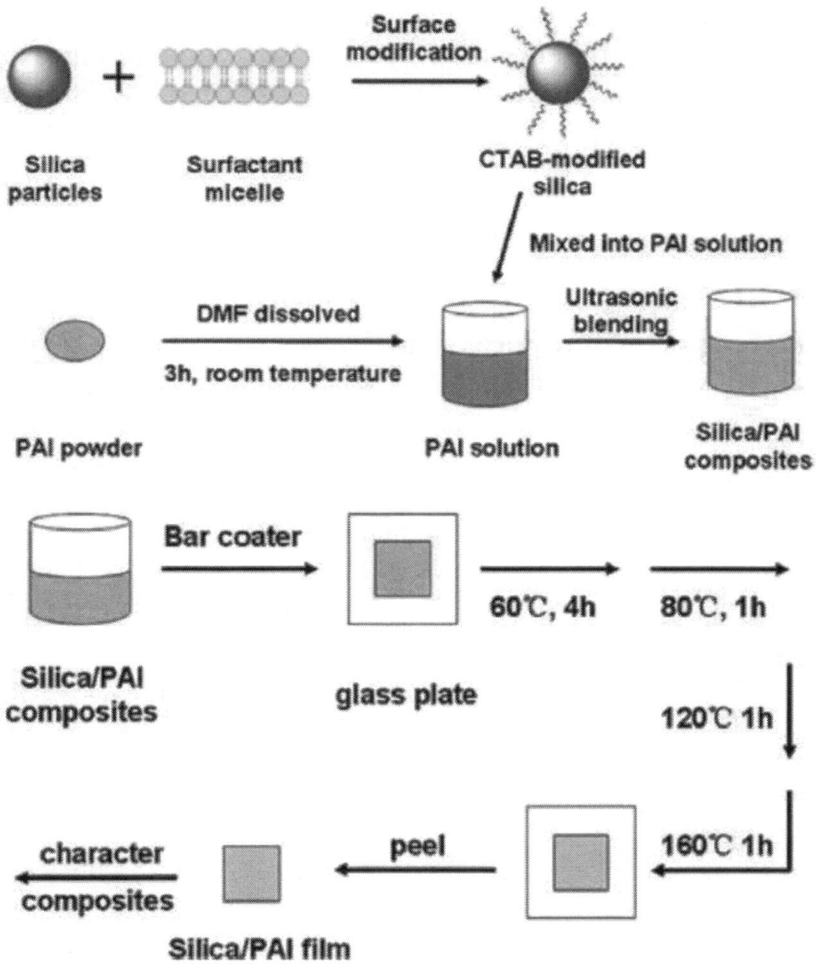

Scheme 1 Preparation process of the silica/PAI nanocomposite films

2.4 Characterization

The fracture surface of composites films were determined by a Hitachi Co. S-4700 scanning electron microscope (SEM). Before SEM imaging, the samples were sputtered with thin layers of Pt–Pd. A Scinco STA S-1500 simultaneous thermal analyzer was applied to analyze the thermal stability of the nano-silica/PAI composite films. The samples were heated from 30 to 800°C at 10°C/min in an air atmosphere. The coefficients of thermal expansion (CET) of silica/PAI composites films were evaluated by a Q 400 EM (USA) thermomechanical analysis (TMA) (5°C/min from 25 to 300°C, 50 mN). All the samples were of 3 × 16 mm, which were cut from original films using a razor blade.

3. RESULTS AND DISCUSSION

3.1 FT-IR Spectra of the Silica/PAI Composite Films

The FT-IR spectra of silica/PAI composite films were shown in Fig. 1. In order to compare the effect of surfactant on composite films, the amount of silica nanoparticles to PAI was all 6 wt%, while the dosage of CTAB to silica was changed from 0 to 3 wt%. The character vibrations of Si–O were found at 1,086, 945 and 796 cm^{-1}, as shown in Fig. 1e. After the surface modification process, the typical stretching vibrations of the C–H were found at 2,855 and 2,928 cm^{-1}, which results from the $-CH_2$ and $-CH_3$in the CTAB. As shown in Fig. 1, the typical characteristic bands of PAI polymer matrix were found. Such as the N–H stretching band at 3,317 cm^{-1}, amide C=O region around 1,710 cm^{-1} and the bands at 1,771 and 1,710 cm^{-1} are associated with the imide carbonyl band. These bands were similar with each other due to the same amount of silica added into the composites. It was worth pointing out that the characteristic stretching vibration of Si–O at 1,086 cm^{-1} became wider, when the silica nanoparticles were modified by CTAB. It could be explained that the organic chain-side of CTAB were grafted on the surface of silica nanoparticles, which made the interaction between the silica nanoparticles and PAI polymer matrix had been improved.

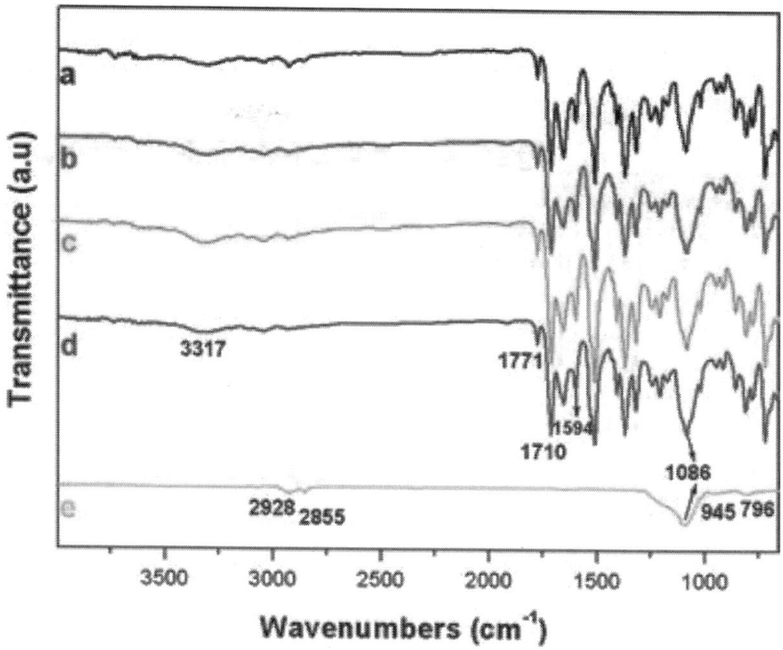

Figure 1 FT-IR spectra of **a** unmodified-silica/PAI composites; **b** 1.0 wt%CTAB-silica/PAI composites; **c** 2.0 wt%CTAB-silica/PAI composites; **d**3.0 wt%CTAB-silica/PAI composites; **e** CTAB-modified silica nanoparticles

Then, the amount of silica to PAI was changed from 2 to 10 wt% and the dosage of CTAB to silica was at 3 wt%, the FT-IR spectra of pure PAI and some silica/PAI composite films were shown in Fig. 2. In those spectra, the bands at 1,771 and 1,710 cm^{-1} were associated with the imide carbonyl band and they were so insensitive to the presence of the silica nanoparticles. The bands in the region from 945 to 650 cm^{-1} increased with the silica content increasing, due to the presence of a broad band associated with the vibration of Si–O band. On the other hands, the N–H stretching band at 3,317 cm^{-1} increased slightly in intensity with the silica content increasing, suggesting hydrogen-bonded N–H groups in the PAI polymer and the Si–O–Si or Si–O–H groups of the silica nanoparticles. The characteristic band at 1,594 cm^{-1} came from the benzene ring stretch and a contribution from O–H bonding in monomeric water, which also has a band at 1,663 cm^{-1}. In conclusion, all the characteristic peaks in the composites indicated that the interaction between the silica nanoparticles and PAI polymer matrix were sensitive to the amount of silica in composites.

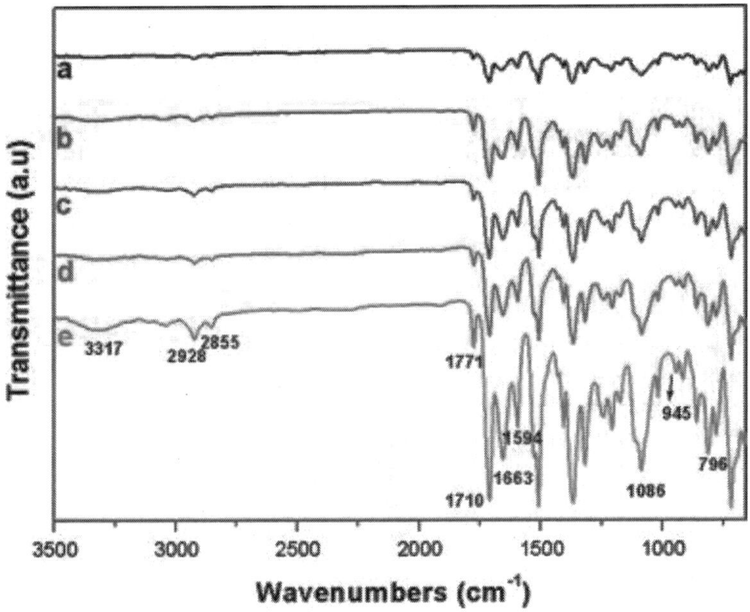

Figure 2 FT-IR spectra of silica/PAI composite with the amount of silica to PAI **a** 2 wt%; **b** 4 wt%; **c** 6 wt%; **d** 8 wt%; **e** 10 wt%

3.2 SEM Micrographs of Silica/PAI Composite Films Fracture Surface

The fracture surface micrographs of silica/PAI composite films are shown in Fig. 3. In Fig. 3a, some silica agglomerations could be found in the micrograph, when the unmodified-silica nanoparticles were added into the PAI matrix. As we knew, silica nanoparticles were easy to agglomerate together; however, the CTAB-modified silica nanoparticles could be dispersed well in the PAI matrix with the amount of CTAB increasing. The dispersal state of the silica nanoparticles was improved when the silica nanoparticles were modified with 1wt % CTAB, as shown in Fig. 3b. In Fig. 3c, the silica nanoparticles were almost monodispersed and a little agglomeration could be found, when 2 wt% CTAB modified silica nanoparticles were added into the PAI polymer. Especially, after the silica were modified with 3 wt% CTAB, the silica nanoparticles became monodispersed without any agglomerations even though the amount of silica nanoparticles to PAI increased to 6 wt%, as shown in Fig. 3d. As a result, the CTAB could improve the dispersal state of silica nanoparticles in a PAI polymer matrix and the optimal dosage of CTAB to silica nanoparticles was 3 wt%.

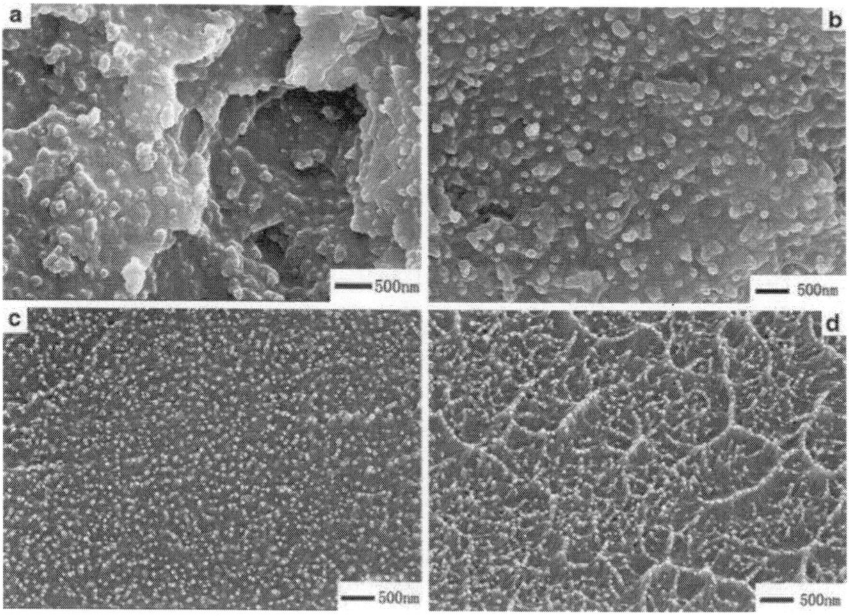

Figure 3 Fracture surface micrographs of **a** unmodified-silica/PAI composites; **b** 1.0 wt%CTAB-silica/PAI composites; **c** 2.0 wt%CTAB-silica/PAI composites; **d** 3.0 wt%CTAB-silica/PAI composites

The fractured surfaces micrographs of pure PAI and several composite films are shown in Fig. 4. First of all, the fracture surfaces of pure PAI film were uniform and the continuous polymer phase is shown in Fig. 4a. Figure 4b–4f shows the fracture surface of different silica/PAI composites, and the amount of silica to PAI changed from 2 to 10 wt%. The more silica nanoparticles were added into PAI, the more of them that could be found in the fracture surface micrographs. In Fig. 4f, when the 10 wt% silica nanoparticles were added into PAI, the silica nanoparticles still were monodispersed without any agglomerations. The results indicated that the CTAB-modified silica nanoparticles had the better dispersal state in the PAI polymer matrix and could increase the content of silica nanoparticles in composites. The surface modification process is an effective method to prepare the silica/PAI nanocomposite.

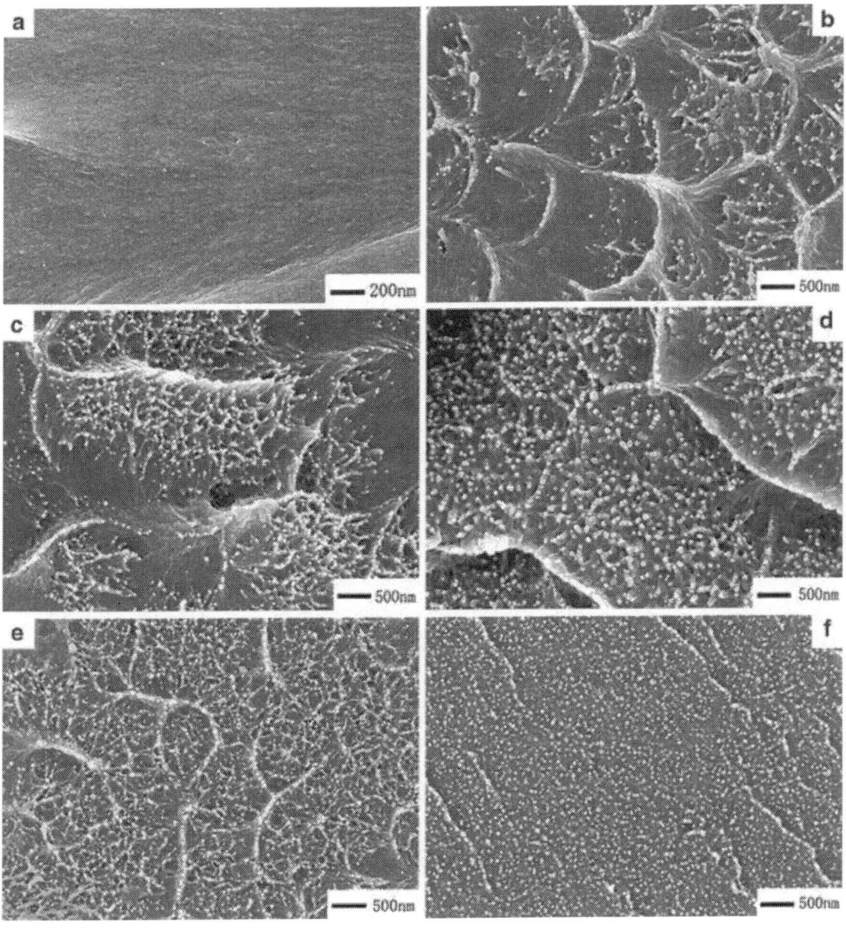

Figure 4 Fracture surface micrographs of **a** pure PAI and silica/PAI nanocomposite with the amount of silica to PAI, **b** 2 wt%, **c** 4 wt%, **d** 6 wt%, **e**8 wt%, **f** 10 wt%

3.3 Thermal Stability of Silica/PAI Composite Films

The thermal stability of silica/PAI composite films was evaluated by thermogravimetric analysis. TGA plots of PAI and composites with the different amounts of silica nanoparticles were shown in Fig. 5. The amount of silica nanoparticles to PAI polymer matrix was changed from 0 to 10 wt% and the plots were listed in Fig. 5a–5f, respectively. A weight loss is observed above 170°C on all the TGA plots, which corresponds to the loss of water and solvent. When the temperature was raised to 450°C, the PAI matrix began to decompose and the decomposition temperature became higher when the silica nanoparticles were added into the PAI polymer matrix. At the same temperature, all the curves of the composites indicated that the weight loss of the composites was less than that of the pure PAI matrix. In other words, the silica/PAI composites have a higher decomposition temperature when the PAI polymer matrix loses the same weight. It is worth pointing out that the thermal stability of PAI was enhanced with the increasing silica contents. In addition, the amount of silica nanoparticles in the composites could be calculated accurately in the TGA plots. Figure 5b–5f shows that the amounts of silica in composites were 1.9, 3.6, 5.8, 7.3 and 10.2 wt%, respectively.

3.4 Coefficient of Thermal Expansion (CTE) of Silica/PAI Composite Films

The coefficient of thermal expansion (CTE) is an important parameter to evaluate the properties of enameled wire. The low CTE could reduce thermal stress build-up and avoid device failure through peeling and cracking at the interface between the polymer film and copper. The CTE curves of pure PAI film and some composite films were shown in Fig. 6. The CTE value of PAI films was 3.87×10^{-5} m/m °C, while the CTE value of silica/PAI composite film was decreased to 3.69×10^{-5} m/m °C and 3.51×10^{-5}m/m °C, when the amount of silica to PAI was 4 and 6 wt%, respectively. The CTE values continuously decreased with the amount of silica particles increasing, especially, the CTE value decreased to 3.35×10^{-5} m/m °C when the amount of silica to PAI was 10 wt%, as shown in

Fig. 6d. Compared with PAI polymer film, the silica/PAI composite films had the lower coefficient of thermal expansion (CTE), due to the rigidity and stiffness of silica nanoparticles and the interaction between the silica and PAI polymer matrix. In conclusion, the rigidity and stiffness of silica nanoparticles limited the movement of the polymer chain, which made the thermal expansion of PAI matrix decrease.

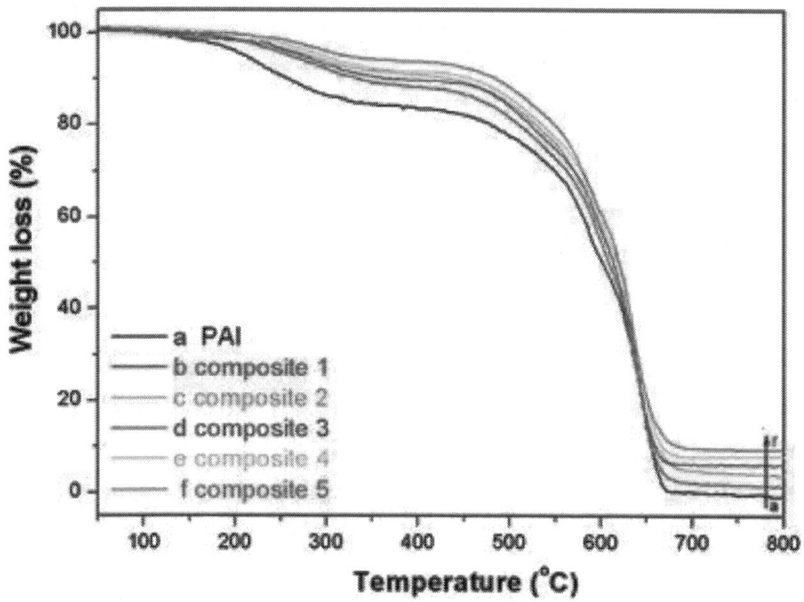

Figure 5 TGA plots of the PAI matrix (**a**) and silica/PAI nanocomposite films with the amount of silica to PAI (**b**) 2 wt%, **c** 4 wt%, **d** 6 wt%, **e** 8 wt%, **f** 10 wt%

4. CONCLUSION

The silica/PAI inorganic/organic nanocomposite films were successfully synthesized by the simple ultrasonic blending, after the silica nanoparticles were modified by CTAB. In the fracture surface micrographs of composite films, the silica nanoparticles were found well dispersed in the PAI polymer matrix. The silica nanoparticles were still monodispersed without any agglomerations when the amount of silica nanoparticles to PAI reached 10 wt%. The thermal stability of PAI was improved, and the decomposition temperature was increased when the amount of silica nanoparticles was increased. The lower coefficient of thermal expansion of composite films could reduce the peeling and cracking at the interface between the polymer film and copper. In this system, the high thermal stability and low coefficient of thermal expansion showed that the silica/PAI nanocomposite films can be widely used in the enamel wire industry.

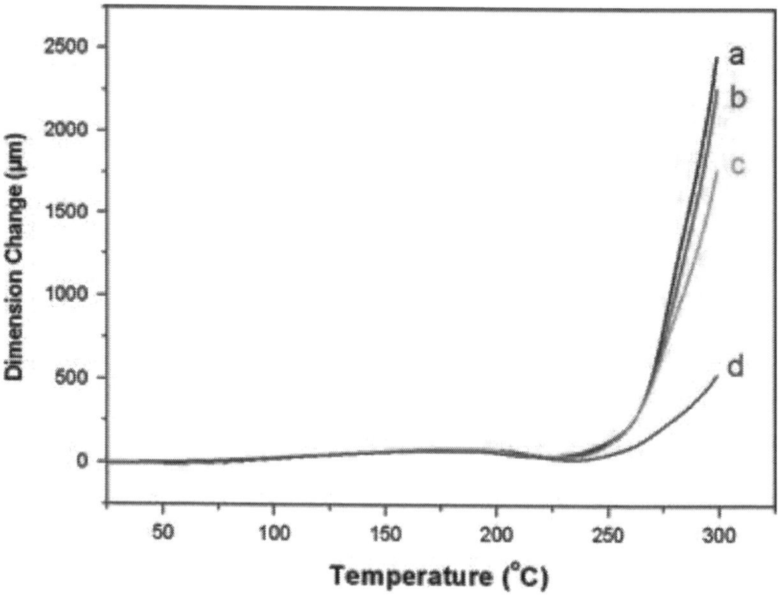

Figure 6 The CTE curves of the **a** pure PAI film; **b** composite films with 4 wt% silica, **c** composite films with 6 wt% silica; **d** composite films with 10 wt% silica

REFERENCES

1. D. Liaw, P. Hsu, B. Liaw, J. Polym. Sci. Pol. Chem. 39, 63 (2001)

2. K. Babooram, B. Francis, R. Bissessur, R. Narain, Compos. Sci. Technol. 68, 617 (2008)

3. D. Liaw, B. Liaw, Polymer 42, 839 (2001)

4. S. Hsiao, C. Yang, Polym. Sci. Pol. Chem. 28, 1149 (1990)

5. D. Liaw, W. Chen, Polym. Degrad. Stabil. 91, 1731 (2006)

6. M. Barikani, S.M. Ataei, J. Polym. Sci. Pol. Chem. 37, 2245 (1999)

7. Y. Wang, S. Goh, T. Chung, P. Na, J. Membrane Sci. 326, 222 (2009)

8. G. Entura, E. Gottardi, I. Peroni, A. Peruzzi, G. Ponti, Nucl. Phys.B 78, 571 (1999)

9. L.W. Chen, K.S. Ho, J. Polym. Sci. Pol. Chem. 35, 1711 (1997)

10. A. Ranade, N. Souza, B. Gnade, Polymer 43, 3759 (2002)

11. H. Kikuchi, Y. Yukimon, S. Itonaga, Hitachi Cable Rev. 21, 55 (2002)

12. K. Kusakabe, K. Ichiki, J. Hayashi, H. Maeda, S. Morooka, J. Membrane Sci. 115, 65 (1996)

13. S.S. Hossein, Y. Lia, T.S. Chunga, Y. Liu, J. Membrane Sci. 302, 207 (2007)

14. M. Castellano, L. Conzatti, G. Costa, L. Falqui, A. Turturro, B. Valenti, F. Negroni, Polymer 46, 695 (2005)

15. M. Alexandre, P. Dubois, Mat. Sci. Eng. R 28, 1 (2000)

16. I.A. David, G.W. Scherer, Chem. Mater. 7, 1957 (1995)

17. Y. Yang, P. Wang, Polymer 47, 2683 (2006)

18. M.D. Butterworth, R. Corradi, J. Johal, S.F. Lascelles, S. Maeda, S.P. Armes, J. Colloid Interf. Sci. 174, 510 (1995)

19. Y. Ouabbas, A. Chamayou, L. Galet, M. Baron, G. Thomas, P. Grosseau, B. Guilhot, Powder Technol. 190, 200 (2009)

20. Y.L. Lee, Z.C. Du, W.X. Lin, Y.M. Yang, J. Colloid Interf. Sci. 296, 233 (2006)

21. S.D. Bhagat, Y.H. Kim, K.H. Suh, Y.S. Ahn, J.G. Yeo, J.H. Han, Micropor. Mesopor. Mater. 112, 504 (2008)

22. C. Oh, Y.G. Lee, C.U. Jon, S.G. Oh, Colloid. Surf. A 337, 208 (2009)

23. L. Xue, J. Li, J. Fu, Y. Han, Colloid. Surf. A 338, 15 (2009)

24. G.L. Jadav, P.S. Singh, J. Membrane Sci. 328, 257 (2009)

25. C. Zhang, M. Zhang, H. Cao, Z. Zhang, Z. Wang, L. Gao, M. Ding, Compos. Sci. Technol. 67, 380 (2007)

2

TEMPERATURE- AND THICKNESS-DEPENDENT ELASTIC MODULI OF POLYMER THIN FILMS

Zhimin Ao and Sean Li

ABSTRACT

The mechanical properties of polymer ultrathin films are usually different from those of their counterparts in bulk. Understanding the effect of thickness on the mechanical properties of these films is crucial for their applications. However, it is a great challenge to measure their elastic modulus experimentally with *in situ* heating. In this study, a thermodynamic model for temperature- (T) and thickness (h)-dependent elastic moduli of polymer thin films $E_f(T,h)$ is developed with verification by the reported experimental data on polystyrene (PS) thin films. For the PS thin films on a passivated substrate, $E_f(T,h)$ decreases with the decreasing film thickness, when h is less than 60 nm at ambient temperature. However, the onset thickness (h^*), at which thickness $E_f(T,h)$ deviates from the bulk value, can be modulated by T. h^* becomes larger at higher T because of the depression of the quenching depth, which determines the thickness of the surface layer δ.

1. INTRODUCTION

As devices are being developed with a view towards making them smaller, thinner and lighter in dimension, thin polymer films are found to be in more stringent demand in various applications, such as diffusion barriers, dielectric coatings, electronic packing, and so on [1]. Therefore, understanding the elastic modulus in confined geometries, such as in thin films, is critical to the stability of the structures of the actual devices. A growing demand exists for the determination of the mechanical properties of thin films and coatings at a rapid pace. Recent researches primarily focusing on the confinement effect of the glass-transition temperatureT_g in thin films [2–7], have presented inconsistent results. It is believed that such a phenomenon might be attributed to the surface and interfacial effects. However, despite its technological importance, the corresponding elastic modulus of the confined thin polymer films has yet to be fully characterized to the same extent as T_g effects have been done.

Direct experimental measures of the ultrathin material modulus have proven to be difficult since the presence of stiff substrate tends to interfere with the measurements [8]. With regard to the impact of confinement on the elastic modulus of soft materials, there is no agreement. This occurrence was caused by the various measurement strategies [9], similar to the initial polymer thin film T_g measurements [4, 6]. The different trends of the elastic modulus are due to the different interfacial interactions between probe and the polymer surface. Thus, noncontact mechanics measurement, which does not disturb the free surface, appears to be potentially advantageous for determining the modulus of polymer films. Recently, a wrinkle-based metrology was developed to measure the elastic properties of thin polymer films [10]. In surface wrinkling measurements, to determine the modulus of thin polymer films, a wrinkling instability is utilized to induce compression of a stiff film bonded to a compliant substrate. The film modulus E_f is determined based on the formula relating the substrate modulus E_{sub}, the film thickness h, and the wrinkling wavelength λ: $E_f = 3E_{sub}\left(\frac{\lambda}{2\pi h}\right)^3$. At room temperature, the deduced elastic modulus of the films decreases with decreasing film thickness for ultrathin polymer films (thickness less than 30 nm) [11]. In order to understand better the physical nature of the thickness dependence of the deduced elastic modulus in ultrathin films, a bilayer model was proposed to account for the surface effect on the wrinkling associated with the surface of a soft layer [12].

In the most recent research, the elastic moduli of a series of poly(methacrylate) films with widely varying bulk glass-transition temperature (T_{gb}) as function of thickness at ambient temperature were measured by wrinkle-based metrology. A decrease of the modulus was found in all ultrathin polymers films (< 30 nm) with the onset of confinement effects shifting to larger film thicknesses as the quench depth into glass state (T_{gb} - T) decreases, where T is the measured temperature [8]. In other words, the quench depth affects the extent of the size dependence. To have a better clarification of the quench-depth effect on the elastic modulus of thin polymer films and the nature of the glass transition of polymers, T was considered as a variant to obtain the temperature-dependent elastic modulus of thin films. However, the *in situ* heating of a polymer-substrate system covering all the processing and characterization steps is impractical [8]. Therefore, in this study, a model will be developed to investigate the temperature- and thickness-dependent elastic moduli of thin polymer films based on both the bilayer and the size-dependent glass-transition temperature thermodynamic models. The results of this new proposed model are then verified with experimental data that were obtained by wrinkle-based metrology for thin polystyrene (PS) films with different molar weights (M_w).

2. THEORETICAL MODEL

It is well known that the atomic or the molecular structure on a surface is different from the bulk structure in solids. As a result, many materials' properties (e.g., mass density, electrical conductivity, elastic modulus, etc.) on the surface differ from their bulk counterparts. Such a difference is negligible for large-scaled structures. For nanostructures, however, the surface-to-volume ratio is large, and the surface effects can be significant. For thin polymer films of interest in this study, one may assume that a surface layer exists with different elastic moduli. The thickness of the surface layer δ may vary from say one atomic layer for crystalline materials [13–15] to a few nanometers for polymers [16, 17]. In the surface layer, lesser density, larger mobility, and softening were found for polymer films [5, 17, 18]. Therefore, a bilayer model was proposed to describe the elastic modulus of thin polymer films E_f [12]. In brief, the model consists of a polymer film with thickness h, containing two distinct moduli. The surface layer of the film has a modulus E_{sur} and a finite depth of δ underneath the atmosphere/solid interface. This depth was considered to be independent of the film thickness. The remainder of the film (h-δ) exhibits a bulk-like modulus (E_{bulk}). The bilayer model can be written as [12]

$$E_f(h) = E_{bulk}\left(1 - \frac{\delta}{h}\right) + E_{sur}\left(\frac{\delta}{h}\right). \quad (1)$$

E_{sur} was considered a few orders of magnitude smaller than the corresponding E_{bulk} [11, 12]. On the other hand, it is known that temperature influences the elastic properties of solids [13, 19]. In general, the elastic moduli of solids would decrease at high temperature due to the weakening of the interactions between atoms or molecules induced by the thermo-expansion of solids [14]. However, Equation 1 ignores the temperature effect, although the mechanical properties of amorphous polymers are slightly influenced by temperature in the glass region. The mechanical properties would change significantly as the temperature approaches the T_g [20]. In addition, the bilayer model considered that δ is independent of h. However, computer simulations have found that a thinner polymer film has a thicker surface soft layer at a given temperature, and the soft layer would extend to almost the whole film at T_g [20]. Recent research has shown that the following relationship exists for bulk polymers: $\delta = \frac{1}{0.005(T_{gb} - T)}$ [8]. Thus, δ is related to the quench depth (T_{gb} - T). In the nanometer-scale range, the relationship is considered to be still valid since polymer films and corresponding bulk polymer have similar thermodynamic behaviors [18, 20]. In addition, simulations have noted that near T_g the thickness of the free surface layer can nearly extend throughout the thin film [20]. Therefore,

$$\delta = \frac{1}{0.005\left[T_g(h) - T\right]}. \quad (2)$$

The glass-transition temperature of thin polymer films $T_g(h)$ is also dependent on the thickness [2–7]. Therefore, after considering temperature effects for $T < T_g(h)$, the bilayer model can be modified as

$$E_f(T,h) = E_{bulk}(T)\left[1 - \frac{1}{0.005h\,(T_g(h) - T)}\right] + E_{sur}\left[\frac{1}{0.005h\,(T_g(h) - T)}\right]. \quad (3)$$

It is noted that the temperature effect on the film thickness is ignored in Equation 3 as the thermal expansion is relatively small compared with the thickness variation here. In this case, $T_g(h)$ can be determined by [4, 5]

$$T_g(h) = T_{gb}\left[\exp\left(-\frac{\alpha_s - 1}{h/h_0 - 1}\right) + \exp\left(-\frac{\alpha_i - 1}{h/h_0 - 1}\right)\right] \Big/ 2 \quad (4)$$

where $h_0 = 2c\,\xi$, with ξ being the correlation length for the intermolecular cooperative rearrangement; c is a parameter related with the surface and interface: $c = 1$ for free-standing thin films or supported thin film with strong interaction between the polymer and the substrate, such as hydrogen bonding; and $c = 1/2$ for a supported thin film with weak interaction between the polymer and the substrate, such as van der Waals force, which is equivalent to the disappearance of the interface. $\alpha_s = [2\Delta C_{pb}/3R]+1$, where$R$ is the ideal gas constant, and ΔC_{pb} is the heat-capacity difference between the bulk glass and the bulk liquid at T_{gb}. $\alpha_i = \alpha_s E_s/E_i$ with E_s and E_i being the bond strength at the surface and interface, respectively.

3. RESULTS AND DISCUSSION

The theoretical model was applied to the PS thin films to verify the newly developed temperature- and thickness-dependent elastic modulus model. First, $E_{bulk}(T)$ and $T_g(h)$ should be determined for the PS films. It is known that the elastic properties decrease almost linearly in the glass state. They present a very strong temperature-dependent behavior near T_g [8, 19]. In addition, generally the glass transition occurs from T_g-50 K for bulk polymers. Therefore, $E_{bulk}(T)$ for bulk PS at $T < T_g$-50 K is a linear function, which can be deduced from the experimental data obtained from $E_{bulk}(T) = -0.00189T + 4.558$ [19]. The result is also in agreement with another experimental result where the elastic modulus of bulk PS, $E_{bulk} = 4.0$ GPa at $T = 294 \pm 3$ K [12].E_{sur} was considered much smaller than the corresponding E_{bulk}; it is about 0.1 GPa for PS [12].

Figure 1 shows the thickness dependence of the elastic modulus of PS thin films at T = 295 K that was obtained from Equation 3using expression (4). The parameters needed for the PS thin films are given in the caption of Figure 1. In this figure, our results are compared with the corresponding experimental results for two different molar weights. These results are well in agreement with each other. The modulus for thick films (> 100 nm) was found to be independent of the film thickness, whereas the elastic modulus decreases with the film thickness when the thickness is less than 60 nm. A similar thickness-dependent behavior was also found for several other polymer films, such as poly(methyl methacrylate) (PMMA), poly(ethyl methacrylate) (PEMA), and poly(isobutyl methacrylate) PiBMA [8, 11]. The depression of the elastic modulus for thin films is a consequence of the soft surface layer, whose relative importance increases as the surface-to-volume ratio increases.

Using Equation 3, the temperature dependence of elastic modulus of PS thin films with h = 10, 30, and 100 nm and bulk are shown in Figure 2. It is noted that the elastic modulus decreases as T increases for all the films. At low temperatures, there is a nearly linear relationship between the elastic modulus and temperature. However, the elastic modulus decreases steeply at the onset of approximately 210, 300, and 350 K for 10, 30, and 100 nm PS thin films, respectively, while the glass-transition temperatures are 275, 347, and 368 K for 10, 30, and 100 nm, respectively, based on Equation 4. The differences between the onset point and T_g are 65, 47, and 18 K, respectively, for the films of three different thicknesses, which imply that the modulus decreases greatly in the glass-transition temperature region, and this region is more extended as the thickness decreases, consistent with the available experimental data in published literatures [19, 21, 22].

Most recently, the deviation of the elastic modulus from its bulk value was studied for thin glassy polymer films with different glass-transition temperatures at ambient temperature. These results suggested that the deviations are significantly influenced by the quench depth into the glass (T_{gb} - T) [8]. To induce the different quench depths, the temperature T is varied in this study. Figure 3 plots the thickness-dependent elastic moduli of PS thin films at different temperatures. For clearly demonstrating the size effect, a relative value of $E(T,h)/E(T,\infty)$ is taken as the ordinate in the figure. From Figure 3, it is found that the size effect is more significant at high temperatures. Recent research has reported that the elastic modulus and glass-transition temperature deviate from the corresponding bulk values at the same thicknesses for poly(*n*-propyl methacrylate) (PnPMA) thin film on poly (dimethylsiloxane) (PDMS) substrate [8]. In Figure 3, $T_g(h)/T_{gb}$ for PS thin films on a passivated substrate is shown by the dashed curve, which is obtained by Equation 4. We find that the functions $T_g(h)/T_{gb}$ and $E_f(150,h)/E(150,\infty)$ behave in a similar manner as a function of h. Also, the thicknesses at which the two functions deviate from the corresponding bulk values are almost the same, as in the case of the PnPMA/PDMS system. Considering a criterion that the deviation of the modulus starts when$E(T,h)/E(T,\infty) \approx 0.96$, the critical film thickness h^*, at which the deviation starts for different temperature, is shown in Figure 4. Note that h^* increases as T increases. In other words, at high temperatures, the size effect is more important, and it can be tuned by the application temperature. Therefore, in actual applications, to avoid decreasing the strength of thin films, one should make sure that the film thickness is larger than h^* for a given temperature. The temperature dependence of h^* is the consequence of the depression of quench depth when the temperature is near the glass-transition region.

The effect of the application temperature on the elastic modulus of thin polymer films is seen from the plots of Figure 5, showing the ratio δ/h^* as a function of temperature. It is noted that this ratio is almost a constant, i.e., approximately 0.004. It is known that the size effect is determined by δ/h, where δ is usually considered as independent of h at a given temperature [8, 12]. However, Figure 6 shows the surface thickness of PS films of different thicknesses at 295 K as obtained from Equation 2. Thus, δ depends on h and increases as h decreases. According to Equation 2, it is known that δ is related to $T_g(h)$, which is dependent on size and can be determined by Equation 4. In the case of PS films on the passivated substrate, such as on PDMS, where there is no strong interaction between the polymer films and the substrate, $T_g(h)$ decreases with decreasing h. Therefore, $\delta(h)$ increases as$h$ decreases due to the depression of $T_g(h)$ at a given temperature. On the other hand, the mobility of the polymer film surface layer can be experimentally investigated by nanoparticle embedding [17, 23] and fluorescence methods [7]. Both methods reported that the surface layer has a thickness of several nanometers, which is consistent with the result of Figure 6.

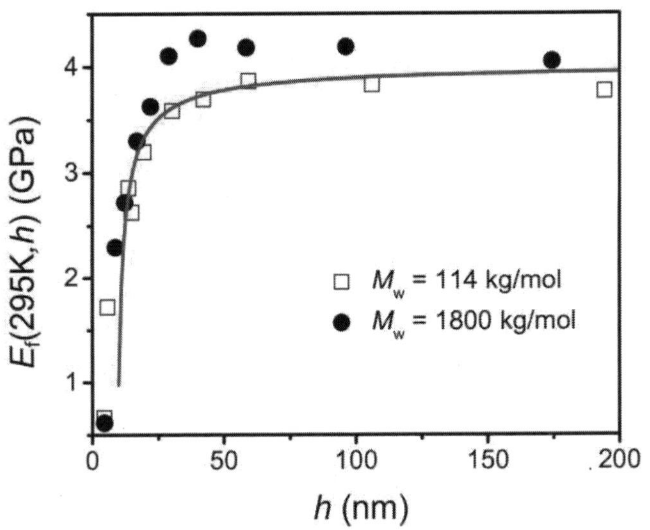

Figure 1 The thickness-dependent elastic modulus of PS thin films at T = 295 K. The symbols circle and square are the experiment results for molar weight M_w = 1800 and 114 kg/mol, respectively [12]. The solid curve is plotted with the calculation results obtained from Equation 3, where the parameters that determine the $T_g(h)$ in Equation 4 are $c = 1/2$, $h_0 = 5$ nm, $\Delta C_{pb} = 30.7$ J mol^{-1} K^{-1} = 1.919 J g atom^{-1} K^{-1}, and $T_{gb} = 375$ K [4].

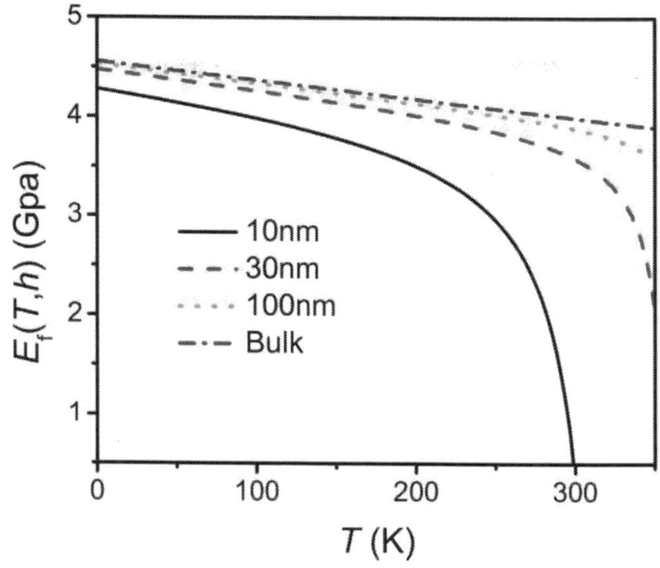

Figure 2 The temperature-dependent elastic modulus of PS thin films with $\boldsymbol{h}$ = 10, 30, and 100 nm and bulk obtained from Equation 3.

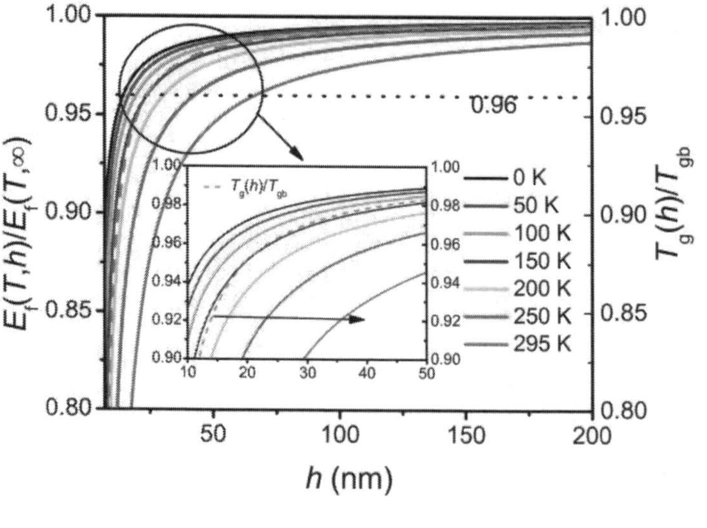

Figure 3 The thickness-dependent elastic modulus of PS thin films at different temperatures. The inset enlarges the encircled region to show the good match of $T_g(h)/T_{gb}$ and $E_f(T,h)/E_f(T,\infty)$.

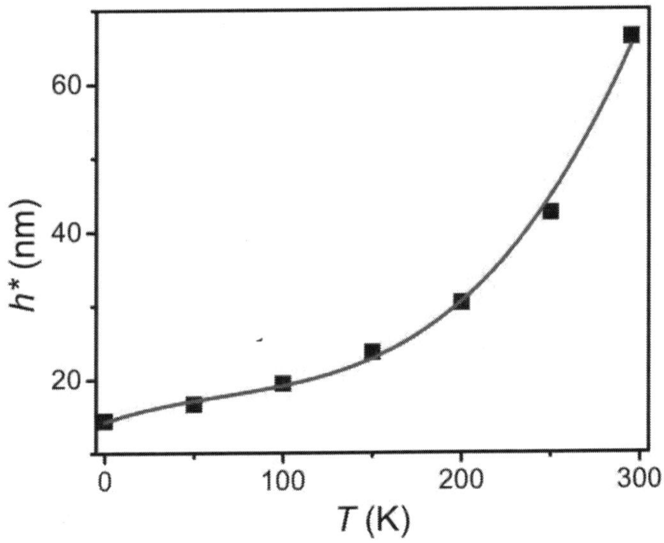

Figure 4 The onset thickness ***h*** * of PS films on the passivated substrate at different temperatures.

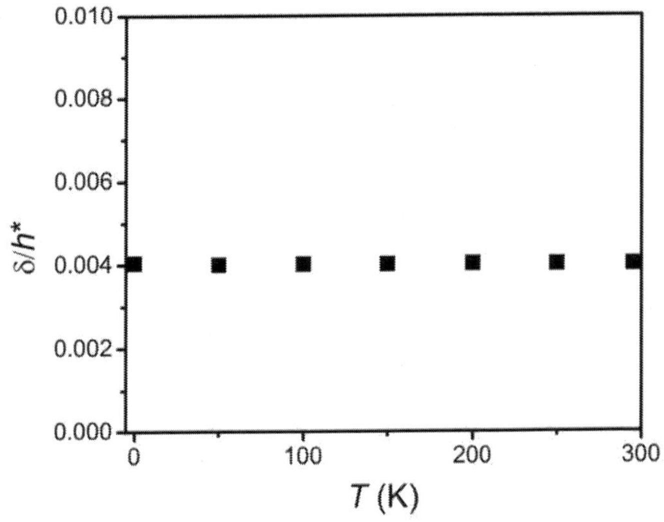

Figure 5 The ratio δ/ ***h*** * as a function of temperature.

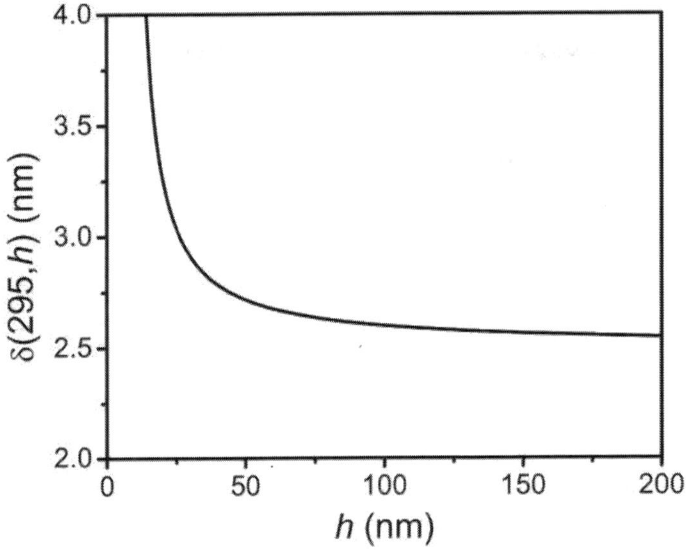

Figure 6 Thickness dependence of the surface layer δ(295 K, ***h***) of PS thin films.

4. CONCLUSION

A theoretical model for the temperature- and thickness-dependent elastic modulus $E_f(T,h)$ was established for amorphous polymer thin films to investigate the dominance of this mechanical property of PS films at nanometer scale. We found that at ambient temperature, $E_f(T,h)$ of PS thin films on the passivated substrate decreases as h decreases when h is thinner than 60 nm, while $E_f(T,h)$ is nearly independent on h for h > 60 nm. Furthermore, a significant thickness effect can be induced by the temperature. The onset of thickness, at which $E_f(T,h)$ deviates from the bulk value, is dependent on temperature and is larger at high temperature. At a certain temperature, $E_f(T,h)$ exhibits the same size-dependent trend as $T_g(h)$, which is associated with the quench depth of $T_g(h) - T$. Except for the temperature effect, the thickness of the surface layer also depends on h, and it increases as h decreases due to the size-dependent glass-transition temperature $T_g(h)$. Therefore, $E_f(T,h)$ of the thin films can be determined using the developed model, thus providing references for the applications of polymer thin films.

ACKNOWLEDGEMENTS

This study was supported by the Vice-Chancellor's Postdoctoral Research Fellowship Program of the University of New South Wales (SIR50/PS19184), the ECR grant of the University of New South Wales (SIR30/PS24201), and the Australian Research Council Discovery Programs DP1096769.

COMPETING INTERESTS

The authors declare that they have no competing interests.

AUTHORS' CONTRIBUTIONS

ZA developed the model and drafted the manuscript. SL co-drafted the manuscript. All authors read and approved the final manuscript.

REFERENCES

1. Elshabini-Riad AAR, Barlow FD: *Thin Film Technology Handbook*. New York: McGraw-Hill; 1997.

2. Keddie JE, Jones RAL, Cory RA: Size-dependent depression of the glass transition temperature in polymer films. *Europhys Lett* 1994, 27: 59–64.

3. Forrest JA, Dalnoki VK: The glass transition in thin polymer films. *Adv Colloid Interface Sci* 2001, 94: 167–195.

4. Jiang Q, Lang XY: Glass transition of low-dimensional polystyrene. *Macromol Rapid Commun* 2004, 25: 825–828.

5. Ao ZM, Jiang Q: Size effects on miscibility and glass transition temperature of binary polymer blend films. *Langmuir* 2006, 22: 1241–1246.

6. Alcoutlabi M, McKenna GB: Effects of confinement on material behavior at the nanometer size scale. *J Phys Condens Matter* 2005, 17: R461-R524.

7. Ellison CJ, Torkelson JM: The distribution of glass-transition temperatures in nanoscopically confined glass formers. *Nat Mater* 2003, 2: 659–700.

8. Torres JM, Stafford CM, Vogt BD: Elastic modulus of amorphous polymer thin filmes: Relationship to the glass transition temperature. *ACS Nano* 2009, 3: 2677–2685.

9. Forrest JA, Dalnoki-Veress K, Dutcher JR: Brillouin light scattering studies of the mechanical properties of thin freely standing polystyrene films. *Phys Phys E* 1998, 58: 6109–6114.

10. Stafford CM, Harrison C, Beers KL, Karim AK, Amis EJ, Vanlandingham MR, Kim H-C, Volksen W, Miller RD, Simonyi EE: A buckling-based metrology for measuring the elastic moduli of polymeric thin films. *Nat Mater* 2004, 3: 545–550.

11. Stafford CM, Vogt BD, Harrison C, Julhongpiput D, Huang R: Elastic moduli of ultrathin amorphous polymer films. *Macromolecules* 2006, 39: 5095–5099.

12. Huang R, Stafford CM, Vogt BD: Effect of surface properties on wrinkling of ultrathin films. *J Aerospace Eng* 2007, 20: 38–44.

13. Ao ZM, Li S, Jiang Q: The determination of Young's modulus in noble metal nanowires. *Appl Phys Lett* 2008, 93: 081905.

14. Sun CQ: Size dependence of nanostructures: impact of bond order deficiency. *Prog Solid State Chem* 2007, 35: 1–159.

15. Ouyang G, Li XL, Tan X, Yang GW: Size-induced strain and stiffness of nanocrystals. *Appl Phys Lett* 2006, 89: 031904.

16. Mansfield KF, Theodorou DN: Molecular dynamics simulation of a glassy polymer surface. *Macromolecules* 1991, 24: 6283–6294.

17. Fakhraai Z, Forrest JA: Measuring the surface dynamics of glassy polymers. *Science* 2008, 319: 600–604.

18. Bohme TR, de Pablo JJ: Evidence for size-dependent mechanical properties from simulations of nanoscopic polymeric structures. *J Chem Phys* 2002, 116: 9939–9951.

19. Fukuhara M: Temperature dependency of elastic moduli and internal dilational and shear frictions of polyetherimide. *J Appl Polym Sci*2003, 90: 759–764.

20. Yoshimoto K, Jain T, Nealey PF, de Pablo JJ: Local dynamic mechanical properties in model free-standing polymer thin films. *J Chem Phys*2005, 122: 144712.

21. Ao ZM, Zheng WT, Jiang Q: Lindemann-like size-independent glass-transition criterion for polymers. *Polymer* 2008, 49: 3578–3581.

22. Tran TA, Saïd S, Grohens Y: Nanoscale characteristic length at the glass transition in confined syndiotactic poly(methyl methacrylate). *Macromolecules* 2005, 38: 3867–3871.

23. Ilton M, Qi D, Forrest JA: Using nanoparticle embedding to probe surface rheology and the length scale of surface mobility in glassy polymers. *Macromolecules* 2009, 42: 6851–6854.

3

IMPROVING THE PERFORMANCES OF RANDOM COPOLYMER BASED ORGANIC SOLAR CELLS BY ADJUSTING THE FILM FEATURES OF ACTIVE LAYERS USING MIXED SOLVENTS

Xiangwei Zhu [1,†], Kun Lu [1,†], Benzheng Xia [1], Jin Fang [1], Yifan Zhao [1], Tianyi Zhao [2,*], Zhixiang Wei [1,*] and Lei Jiang [2,*]

[1] National Center for Nanoscience and Technology, Beijing 100190, China

[2] Key Laboratory of Bio-Inspired Smart Interfacial Science and Technology of Ministry of Education, School of Chemistry and Environment, Beihang University, Beijing 100191, China

ABSTRACT

A novel random copolymer based on donor–acceptor type polymers containing benzodithiophene and dithienosilole as donors and benzothiazole and diketopyrrolopyrrole as acceptors was designed and synthesized by Stille copolymerization, and their optical, electrochemical, charge transport, and photovoltaic properties were investigated. This copolymer with high molecular weight exhibited broad and strong absorption covering the spectra range from 500 to 800 nm with absorption maxima at around 750 nm, which would be very conducive to obtaining large short-circuits current densities. Unlike the general approach using single solvent to prepare the active layer film, mixed solvents were introduced to change the film feature and improve the morphology of the active layer, which lead to a significant improvement of the power conversion efficiency. These results indicate that constructing random copolymer with multiple donor and acceptor monomers and choosing proper mixed solvents to change the characteristics of the film is a very promising way for manufacturing organic solar cells with large current density and high power conversion efficiency.

KEYWORDS

random copolymer; photovoltaic; mixed solvents; film feature

1. INTRODUCTION

A mixed bulk heterojunction (BHJ) structure, which consists of a conjugated polymer as an electron donor and a fullerene derivative as an electron acceptor, produces high power conversion efficiency (PCE) in polymer solar cells (PSCs) [1]. Alternating electron donor (D) and acceptor (A) building blocks to construct D–A type polymers is an effective strategy for designing high efficient photovoltaic materials for PSCs [2,3,4]. The properties of D–A copolymers can be fine-tuned by varying the type of electron-donating or withdrawing unit, respectively. In this way, some highly efficient D–A polymers with improved absorption spectra and proper energy levels have emerged in recent years [5,6].

Current D–A copolymer design strategies include: (1) synthesizing novel donor [7] and acceptor units [8]; (2) introducing π-spacers to construct D-π-A polymers [9]; (3) optimizing the alkyl side chains to balance the crystallinity and solubility [10]; and (4) introducing substituents, for example fluorine atoms, to tune the energy levels [11]. In addition to these methods, introducing many different units into the polymer backbone to construct the copolymer is also a promising approach, since they can exhibit the synergetic effects of each unit by tuning the composition ratio, thus exhibiting red-shifted absorption spectra, proper highest occupied molecular orbital (HOMO) energy levels, higher hole mobility, and greatly improved photovoltaic properties [12]. Such a design strategy creates an intriguing possibility to modify the existing D–A type photovoltaic polymers for better performances. Thus far, there are plenty of examples of significant improvement being achieved by randomly copolymerizing one donor and one acceptor with a third monomer [13,14,15,16].

In addition, other strategies have also been carried out to improve the photovoltaic performances of PSCs by tuning the morphologies of the blend films, such as the selecting different preparation solvent, introducing high boil point additive, using mixed solvents and annealing the active layer [17]. Among them, using mixed solvents is a very maneuverable and efficient way [18]. Since different active layer films prepared by diverse solvents have, respectively, special surface energy, reflecting the surface composition of blend film. Furthermore, the performance of BHJ solar cell in part depends on whether polymer or [6,6]-phenyl-C71-butyric acid methyl ester ($PC_{71}BM$) would be enriched on the top surface of the blend film [19]. Besides, mixed solvents can also be effectively used to tune the miscibility of the donor and acceptor components, the domain size and the morphology of active layer, which play an important role in optimizing the photovoltaic performances for PSCs [20,21,22].

In this study, we adopt an alternative approach for designing copolymer by "doping" a D–A fraction to another D–A copolymer by chemical synthesis. We present a new fully conjugated random copolymer, PSTDPP (see Scheme 1), containing two donor units, siloledithieno (SDT) [23] and benzodithiophene (BDT) [24], and two acceptors, benzothiadiazole (BT) [25] and diketopyrrolopyrrole (DPP) [26]. All of these units are frequently used building blocks for high efficient PSCs [27,28,29,30]. The copolymer was synthesized from them via Palladium-catalyzed Stille coupling reaction. The polymer synthesis concerns the copolymerization of three components, in which BDT acts as a bridge, linking the D–A fraction (BT-SDT-BT) and DPP units. As expected, the random polymer exhibits a broad absorption with a range of 500–800 nm and a proper HOMO energy level, which ensure its applicability for fabrication of PSC devices. The best PCE value of 4.38% was obtained by using mixed solvent of *O*-dichlorobenzene (ODCB) and chloroform (CF) with 1,8-octanedithiol (DIO) as additive, which is much better than those of devices using single CF or ODCB as processing solvent. The result suggests that using mixed solvent to prepare the active layer could improve the film characteristic effectively, leading to an enhanced PCE for PSCs.

2. EXPERIMENTAL SECTION

2.1. Materials and Synthesis

Solvents and other common reagents were obtained from Beijing Chemical Plant (Beijing, China). All other chemicals were purchased from commercial sources (Alfa Aesar (Heysham, UK), Acros (Geel, Belgium) and Sigma-Aldrich (Saint Louis, MO, USA)) and were used without further purification unless otherwise stated. Tetrahydrofuran, toluene, and chloroform were freshly distilled from appropriate drying agents prior to use. Compounds **1**, **2** [23], **8** [31], **10** [32] and **11** [33] were prepared according to the published literature. The detailed synthetic processes are illustrated in Scheme 1.

Compound 3

Compound **2** and 40 mL THF were put into a flask, and cooled down to −78 °C by a liquid nitrogen–acetone bath. Then, 3.9 mL butyllithium solution in hexane (2.7 mol·L^{-1}) was added dropwise for 5 min, and the reactant was stirred for another 15 min at that temperature. Subsequently, 1.95 g dichlorodi(2-ethyl-hexyl)silane (6 mmol) was added in one portion, and the cooling bath was removed and the reactant was stirred for 2 h under ambient temperature. Then, the reactant was poured into water and extracted by ethyl ether several times. After removal of volatiles, the residue was purified by silica gel chromatography using hexane as eluent and gave 2.26 g of compound **3** (yield 72%). The product was obtained as colorless oil. GC–MS: *m*/*z* =

562.2. ^1H NMR ($CDCl_3$, 400 MHz), δ (ppm): 7.06 (s, 2H), 1.71 (m, 2H), 1.51–1.14 (m, 16H), 0.91 (t, 6H), 0.85 (t, 6H), 0.76 (m, 4H), 0.31 (s, 18H).

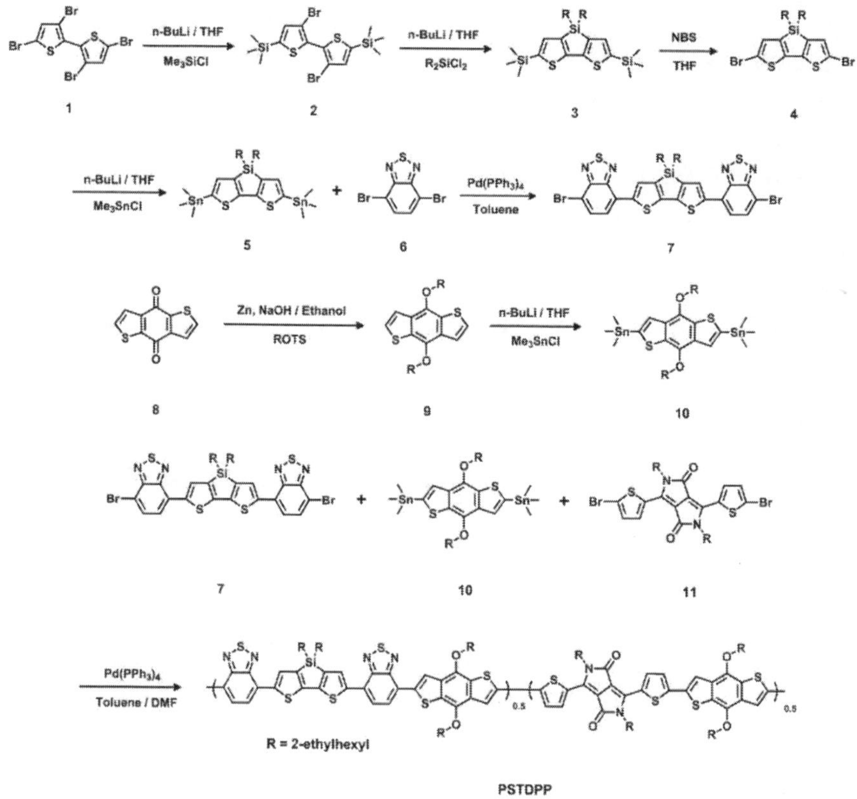

Scheme 1. Synthesis route of the polymer.

Compound 4

Compound **3** (1.69 g, 3 mmol) was dissolved into 20 mL THF, and *N*-bromosuccinimide (1.1 g, 6.17 mmol) was added in one portion. The reactant was stirred at ambient temperature for 4 h, and then extracted by diethyl ether. The volatiles were removed under vacuum, and the residue was purified by silica gel chromatography using hexane as eluent. The product was obtained as sticky colorless or pale yellow oil with a yield of 96%. MS (MALDI-TOF): m/z = 576.0. ^{1}H NMR ($CDCl_3$, 400 MHz), δ (ppm): 6.97 (s, 2H), 1.72 (m, 2H), 1.51–1.13 (m, 16H), 0.91 (t, 6H), 0.83 (t, 6H), 0.76 (m, 4H).

Compound 5

Compound **4** (1.2 g, 2.51 mmol) and 20 mL ultra-dry THF were put into a flask. The clear solution was cooled down to −78 °C by a liquid nitrogen–acetone bath. Then, 2.4 mL butyllithium solution in hexane (2.3 mmol, 2.4 mol/L) was added dropwise. After stirring at −78 °C for 15 min, 7 mL trimethyltin chloride (1 mol/L) was added in one portion, and then the cooling bath was removed. After being stirred at ambient temperature for 2 h, the reactant was poured into cool water and extracted by diethyl ether several times. After removal of volatiles, the product (1.78 g, 2.39 mmol) was obtained as sticky pale green oil with a yield of 95% and used without any further purification. MS (MALDI-TOF): m/z = 744.1. ^{1}H NMR ($CDCl_3$, 400 MHz), δ (ppm): 7.03 (s, 2H), 1.68 (m, 2H), 1.4–1.13 (m, 16H), 0.90 (t, 6H), 0.83 (t, 6H), 0.74 (m, 4H).

Compound 7 [34]

Compound **5** (760 mg, 1.02 mmol), compound **6** (1.8 g, 6.11 mmol), $Pd(PPh_3)_4$ (0.100 g, 0.08 mmol), and toluene (15 mL) were put into a 50 mL flask and stirred under N_2 protection for 20 min. The mixture was heated to 80 °C for 10 min, 100 °C for 10 min and 150 °C for 5 h. Upon cooling, the red residue was passed through a short silica plug eluting with $CHCl_3$ (500 mL). All volatiles were removed *in vacuo* to give the crude product as a dark sticky solid. The material was then slurried in methanol (300 mL), filtered, loaded onto silica using $CHCl_3$ and purified by flash chromatography using a hexanes/$CHCl_3$gradient. The product eluted as a red solution at 40% $CHCl_3$. After fraction collection and solvent removal, the solid product was slurried in methanol (300 mL), sonicated for 30 min, and filtered. The solid was washed with copious amounts of methanol and acetone, and then dried under vacuum for 24 h. The product was collected as green/purple metallic colored powder. Yield = 45%. MS (MALDI-TOF): m/z = 844.0. ^{1}H NMR ($CDCl_3$, 400 MHz), δ (ppm): 8.17 (m, 2H), 7.82 (d, 2H), 7.68 (d, 2H), 1.54 (m, 2H), 1.33 (m, 4H), 1.28 (m, 4H), 1.21 (m, 8H), 1.10 (m, 4H), 0.82 (m, 12H).

PSTDPP

Compound **7** (0.42 g, 0.5 mmol), compound **10** (0.77 g, 1 mmol) and compound **11** (0.34 g, 0.5 mmol) were dissolved into anhydrous toluene (10 mL) in a three-neck flask. The solution was flushed with argon for 20 min, and then $Pd(PPh_3)_4$ (50 mg , 0.043 mmol) was added into the flask. The flask was subjected to three successive cycle of vacuum followed by refilling with argon. Then, the polymerization was carried out at 110 °C for 18 h under argon protection. After cooling to room temperature, the reaction mixture was dripped into 100 mL of methanol with vigorous stirring. The polymer precipitate was collected by filtration and transferred to a thimble. After drying, the crude polymer was subjected to Soxhlet extraction using methanol, acetone, hexane and chloroform as solvents. After final extraction with chloroform, the chloroform solution was concentrated to approximately 10 mL and then dripped into 100 mL of methanol with vigorous stirring. The polymer was collected by filtration and dried to afford a solid as the product. Yield = 72%, M_n = 21.6 kDa and PDI = 3.47.

2.2. Measurements

Mass Spectra were determined on Bruker microflex MALDI-TOF mass spectrometer (Bruker, Billerica, MA, USA). ^{1}H NMR (400 MHz) was obtained on a Bruker avance 400 NMR Spectrometer (Bruker, Billerica, MA, USA) using tetramethylsilane as internal standard. UV–Vis absorption spectra were recorded on a Shimadzu UV-3600 UV–vis–NIR spectrophotometer (Shimadzu, Kyoto, Japan). The electrochemical cyclic voltammetry (CV) was conducted on an EG&G Princeton Applied Research VMP3 workstation (Bio-Logic, Claix, France) with Pt disk coated with the polymer film, Pt plate, and Ag/Ag^+ electrode as working electrode, counter electrode and reference electrode respectively in a 0.1 mol·L^{-1}tetrabutylammoniumhexafluorophosphate (Bu_4NPF_6) acetonitrile solution. The current density–voltage (*J*–*V*) curves were obtained by a Keithley 2420 source-measure unit (Keithley, Cleveland, OH, USA). The photocurrent was measured under illumination using an Oriel Newport 150W solar simulator (AM 1.5G, Newport, Irvine, CA, USA), and the light intensity was calibrated with a Newport reference detector (Oriel PN 91150V, Newport, Irvine, CA, USA). The external quantum efficiency (EQE) measurements of the devices were performed in air with an Oriel Newport system (Model 66902, Newport, Irvine, CA, USA). The thickness of the active layer was measured on a Kla-TencorAlpha-StepD-120 Stylus Profiler (Kla-TencorAlpha-StepD-120 Stylus Profiler). X-ray diffraction (XRD) measurements were carried out in the reflection mode at room temperature using a 12-Kw D/max-rA X-ray diffraction system (12-kW D/max-rA X-ray diffraction system). The step size for all scans was 4 degree with a count time of 1 min per step. Atomic Force Microscope (AFM) images were obtained on a NanoscopeIIIa AFM Multimode Digital Instrument (Bruker, Billerica, MA, USA) in tapping mode.

2.3. Organic Field-Effect Transistor (OFET) Devices Fabrication and Measurement

OFETs were constructed on octadecyltrichlorosilane (OTS) modified SiO_2/Si substrates with top-contact configuration. The 10 mg·mL^{-1} ODCB solutions of the polymers were spin-coated on the substrates in air at room temperature. The source and drain electrodes were patterned through the mask using thermal evaporation method. The channel length were 100 μm and the ratio of channel width to channel length W/L = 110. The mobility in the saturation regime was determined using the equation: $I_{DS} = (\mu_{FET} \cdot W \cdot C_i/2L)(V_G - V_T)^2$, where I_{DS} is the drain-source current in the saturated regime, μ_{FET} is the field-effect mobility, W the channel width, L the channel length, C_i the capacitance SiO_2 dielectric layer, V_G the gate voltage andV_T the threshold voltage.

2.4. PSC Devices Fabrication and Measurement

Polymer solar cell devices with the structure glass/indium tin oxide (ITO)/poly(3,4-ethylenedioxythiophene):poly(styrenesulfonate) (PEDOT:PSS)/polymer:$PC_{71}BM$(w/w)/Ca/Al were fabricated as follows: The PEDOT:PSS solution was spin-coated on top of the cleaned ITO-coated glass substrate and the film thickness was approximately 40 nm. Then the coated substrates were transferred into a glove box. The polymer/$PC_{71}BM$ blend solution used in this study for spin-coating was 10 $mg{\cdot}mL^{-1}$ for the different solutions. The additive, DIO, was added prior to spin-coating process. The thickness of the active layer was controlled by changing the spin speed during the spin-coating process and distributed around 100 nm. The devices were completed by evaporating Ca/Al metal electrodes with an area of 4 mm^2as defined by masks. These layers were thermally evaporated at a pressure of 2×10^{-6} Torr. To optimize device performance, different D/A weight ratios (*w*/*w*) and different solution were used during the device fabrication process.

3. RESULTS AND DISCUSSION

3.1. Optical and Electrochemical Properties

The optical property of PSTDPP was investigated by UV–vis absorption spectra both in chloroform solution and thin films. As shown in Figure 1, the polymer showed a relatively broad absorption in the UV–vis range. The absorption spectrum of PSTDPP in chloroform solution and at solid state showed absorption maximum at 677/738 and 683/747 nm, respectively. Compared with the absorption in solution, a very slight red shift of the spectrum for the solid state was observed. The optical band gap of PSTDPP can be calculated from the film absorption edge as 1.53 eV.

The electrochemical property of the polymer was investigated by cyclic voltammetry (CV) method. As shown in Figure 2, the polymer showed reversible *p*-doping/dedoping (oxidation/re-reduction) and n-doping/ dedoping (reduction/reoxidation) processes. The energy levels of the highest occupied molecular orbital (HOMO) and the lowest unoccupied molecular orbital (LUMO) can be calculated from the onset oxidation and reduction potentials according to the equation: HOMO = $-e\ (E^{ox} + 4.71)$ (eV) and LUMO = $-e\ (E^{red} + 4.71)$ (eV) [35]. The electrochemical band gaps (E_g^{ec}) can be calculated from the equation:$E_g^{ec} = (E^{ox} - E^{red})$

(eV). The HOMO levels and the LUMO levels of PSTDPP were estimated to be at −5.15 eV and −3.47 eV, respectively. The corresponding electrochemical band gap of PSTDPP was 1.68 eV.

3.2. Crystallinity

To gain a deeper insight of the crystallinity and molecular stacking mode of the polymer, X-ray diffraction (XRD) analysis of the polymer film spin-coated on the quartz substrate was operated (Figure 3). The PSTDPP film shows a distinct diffraction peak (100) at 2θ around 4.2°, indicating the chains form a interdigitation packing structure in the film with interchain *d*-spacing distance of 21.0 Å. The peak around 20°–25° for the polymer can be assigned to the face-to-face stacking of the aromatic groups arranged in the parallel direction [36]. The polymer exhibited good crystallinity and formed highly ordered structure in the film, which was expected to afford high hole mobility and facilitate charge carrier transport in BHJ solar cells.

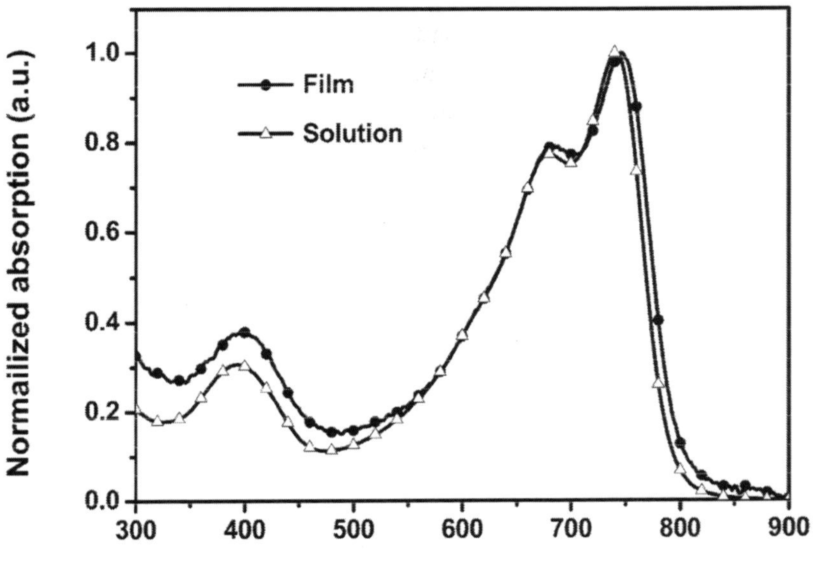

Figure 1. UV–vis absorption spectra of PSTDPP in 1 × 10^{-5} M chloroform solution and at solid state.

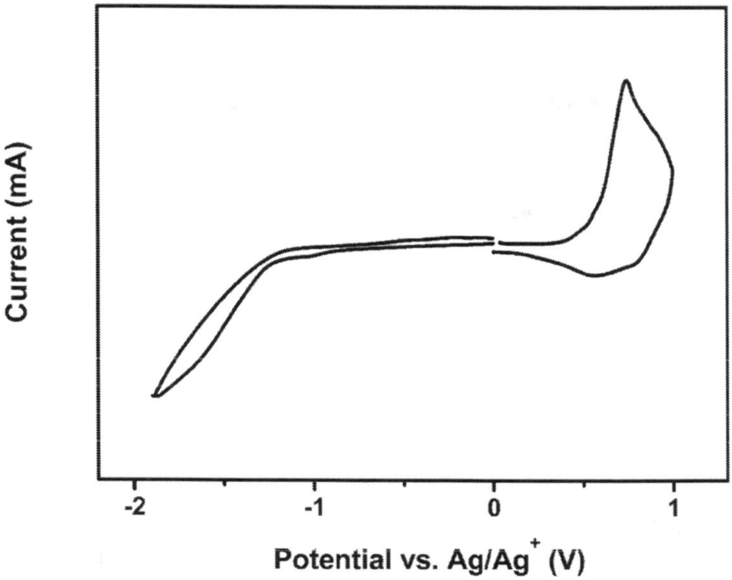

Figure 2. Cyclic voltammetry (CV) curves of PSTDPP at a scan speed of 50 mV s^{-1}.

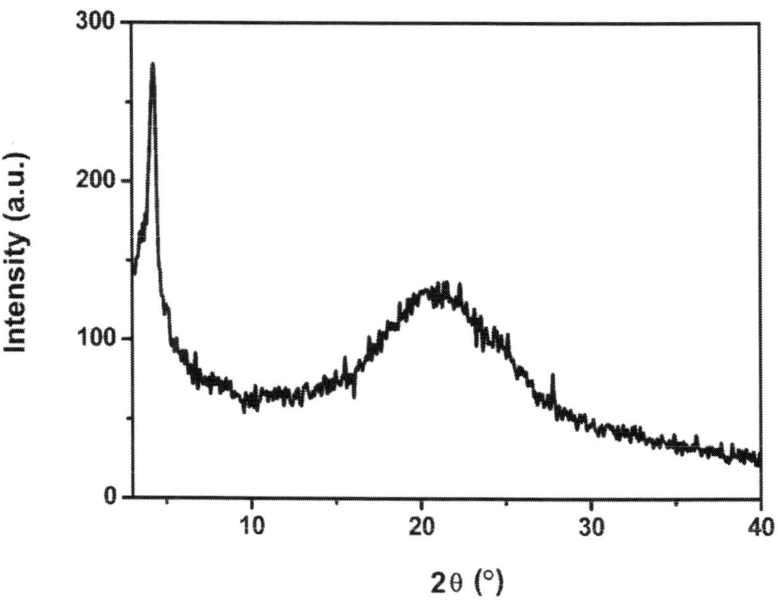

Figure 3. X-ray diffraction (XRD) pattern of PSTDPP film.

3.3. Hole Mobility

The incorporation of siloledithieno and benzodithiophene donor units and strong electron-drawing benzothiadiazole and diketo-pyrrolopyrrole units into the polymer backbones was expected to enhance hole transport mobility of the polymers. The organic field-effect transistors (OFETs) with a conventional bottom-contact and bottom-gate configuration were fabricated using the polymer as the active layer. Figure 4 shows the representative transfer and output current–voltage plots for the OFETs fabricated on an OTS treated Si/SiO_2 substrates. The devices showed a typical *p*-type transporting characteristic measured under ambient conditions. The PSTDPP based OFET exhibited relatively high hole mobility of 2.2×10^{-3} $cm^2 \cdot V^{-1} \cdot s^{-1}$ with an on/off ratio of 10^6. The high mobility of the PSTDPP could be ascribed to the better molecular organization in the film. Furthermore, devices exhibited relatively high stability at ambient conditions. The devices were measured sequentially for about one month, and showed no obvious decline in the performance. The relatively high hole mobility is necessary for efficient charge transport in PSCs, which would be helpful to increase current density (J_{sc}) and fill factors (*FF*) of the devices [37].

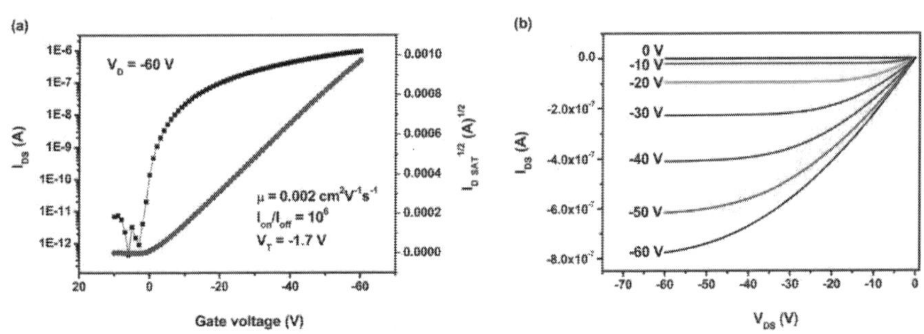

Figure 4. Transfer (**a**) and output (**b**) characteristics of PSTDPP.

3.4. Photovoltaic Properties

Since the polymer possessed proper energy levels and a suitable high mobility, it could be expected to show excellent photovoltaic performance. PSCs were fabricated with a conventional sandwich structure of ITO/PEDOT:PSS/polymer:$PC_{71}BM$/Ca/Al, where the active layer was composed of PSTDPP as the donor and $PC_{71}BM$ as the acceptor. Different D/A ratios (*w/w*) and processing solvents were systematically investigated to obtain the optimal device performance.

The device performances were firstly optimized by changing the D/A ratios of the polymer:$PC_{71}BM$ in ODCB solutions, and the optimal D/A ratio is 1:3 (Table 1). Then, further optimization was carried out by changing the processing solvents (from ODCB to CF or ODCB/CF blend). It was found that using CF as processing solvent was beneficial to the improvement of the open circuit voltage (V_{oc}) and FF. Consequently, the solvent mixtures of ODCB and CF were chosen as processing solvents to investigate whether binary solvent can enhance the overall device performances. When using ODCB/CF (2/1, *v/v*) as processing solvent, a much better PCE of 3.79% was obtained with J_{sc} of 11.13 mA·cm^{-2}, V_{oc} of 0.69 V and *FF* of 49.2%. At this point, all three parameters have been improved significantly compared with using ODCB as processing solvent. Finally, 1,8-diiodooctane (DIO) was added to the blend solution which was expected to tune the film morphology and device performance [38]. The resulting PCE for PSTDPP has grown to 4.38% by adding 2.5% DIO with further improvement ofJ_{sc}/*FF* and minor loss of V_{oc}.

The corresponding V_{oc}, J_{sc}, *FF* and PCE values of the devices are summarized in Table 1. Figure 5a shows the typical current density–voltage (*J–V*) curve of the PSC device fabricated the optimal condition by using the ternary solvents. The accuracy of the photovoltaic measurements for the optimal device can be further confirmed by the EQE of the devices as showed in Figure 5b. The EQE curve exhibited a broader photoresponse in the range of 390 to 750 nm with EQE values of 42%–50%. The J_{sc} calculated from integration of the EQE curve was 11.28 mA·cm^{-2}, which shows around 2.4% mismatch compared with the J_{sc} value obtained from the *J–V* measurement.

Table 1. Photovoltaic properties of the PSTDPP/$PC_{71}BM$-based bulk heterojunction (BHJ) solar cells fabricated with different solvents.

Solvent	D/A (*w/w*)	V_{oc} (V)	J_{sc} (mA cm^{-2})	*FF* (%)	PCE (%)
ODCB	1:1	0.51	5.53	34.8	0.98
ODCB	1:2	0.64	6.75	47.1	2.04
ODCB	1:3	0.63	9.89	37.9	2.35
ODCB	1:4	0.64	8.74	40.9	2.27
CF	1:3	0.69	1.68	50.9	0.59
ODCB:CF (2:1)	1:3	0.69	11.13	49.2	3.79
ODCB:CF (2:1) with 2.5%DIO	1:3	0.68	11.55	55.7	4.38

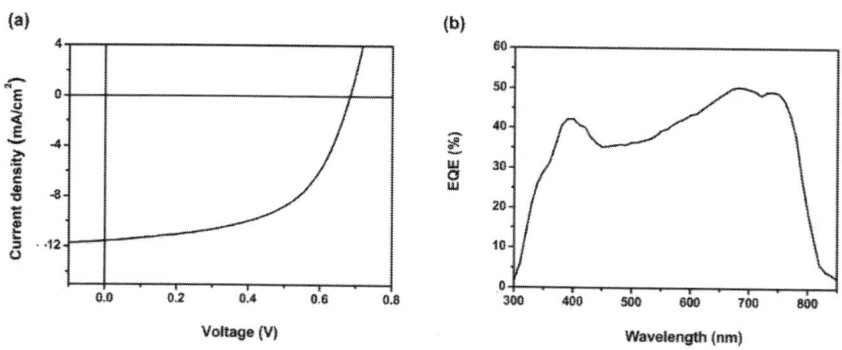

Figure 5. (**a**) *J–V* curve of PSTDPP based BHJ solar cells prepared under the optimal condition (D/A ratio = 1:3, using *O*-dichlorobenzene (ODCB) and chloroform (CF) (*v*/*v*= 2:1) with 2.5% DIO as processing solvent); and (**b**) EQE curve of the corresponding device.

3.5. Water Contact Angle Characterizations

Water contact angle characterizations were used to gain a deep insight of the surface components of the blend films processed by different solutions. As the decrease of water contact angle, the surface energy will increase and thus the top surface of the film will be more hydrophilic. Generally, water contact angle of polymer is bigger than that of $PC_{71}BM$. For a blend film of BHJ solar cell composed by polymer and $PC_{71}BM$, therefore, a smaller water contact angle indicates that $PC_{71}BM$ will be relatively enriched in the top surface of the blend film, and hence favorable ohmic contact with cathode will be formed, thus enhancing the photovoltaic performance of the corresponding device [19]. In fact, water contact angle of the blend film was affected, not only by the characters of polymer and $PC_{71}BM$, but also by the preparation conditions of active layer. Figure 6 shows the respective water contact angles of four active-layer films prepared by different solvents. As shown, the water contact angle of the blend film prepared by CF is 99.4° ± 0.3°, which is far less than that of the blend film prepared by ODCB (105.1° ± 0.3°). Using the mixed solvent of ODCB/CF (2/1, *v*/*v*) to prepare the blend film could decrease the water contact angle and the corresponding value is 103.3° ± 0.5°. Interestingly, when adding some amount of DIO to the mixed solvent, a smaller water contact angle of 100.1° ± 0.4° was observed similar to the blend film prepared by CF, indicating that $PC_{71}BM$ was relatively enriched in the top surface of the blend film in both cases and the corresponding BHJ solar cells might represent relatively high photovoltaic performances accordingly [39,40]. Admittedly, this is just a possibility because the performance of PSC has a lot to do with the morphology of the active layer. Therefore, the analysis of morphology will be discussed in the following section.

3.6. Film Morphologies

Atomic force microscopy (AFM) was used to characterize the morphology of the active layers spin-coated from different solutions. AFM phase image of the blend film prepared by CF (Figure 7a) reveals a high level of aggregation in the active layer and the surface roughness of the film is 0.819 nm. The very clear large domains with sizes of hundreds of nanometers would diminish exciton migration to the D–A interface and influence the following charge separation and collection [41], leading to the worst performance for PSTDPP based PSCs. Using ODCB as the preparation solvent, the morphology of active layer has improved somewhat and the surface roughness becomes a little smoother (root mean square (RMS) roughness = 0.768 nm), but there are still some unequally distributed granular aggregations on the surface of the blend film (Figure 7b). Consequently, the overall performance of the corresponding device showed only moderate improvement. When introducing the mix solvent of ODCB and CF to the active layer preparation, an obvious change of film morphology was seen in the Figure 7c. The granular aggregations have disappeared from the phase image and the domain size has reduced to tens of nanometers. Although the roughness of the film is 0.803 nm, the film morphology is much better than that of the above two cases, so the PCE exhibited a significant improvement. Finally, the optimized morphology was obtained by adding additive (DIO) to the mixed solvent. The roughness of the film has drop to 0.751 nm and the surface of the blend film becomes much smoother (Figure 7d), which is beneficial to the higher efficiency of exciton diffusion as well as charge separation [42]. In addition, combined with the water contact angle result, the best photovoltaic performance was eventually achieved for PSTDPP based PSCs with a ternary solvent.

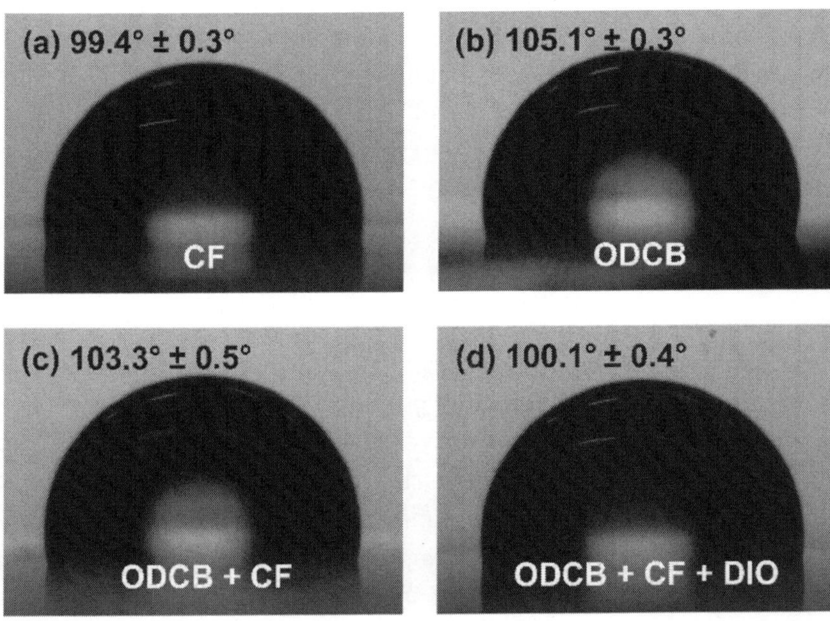

Figure 6. Water contact angle images of the top surface of (**a**) active-layer film prepared by CF; (**b**) active-layer film prepared by ODCB; (**c**) active-layer film prepared by ODCB:CF (2:1); and (**d**) active-layer film prepared by ODCB:CF (2:1) with 2.5% DIO.

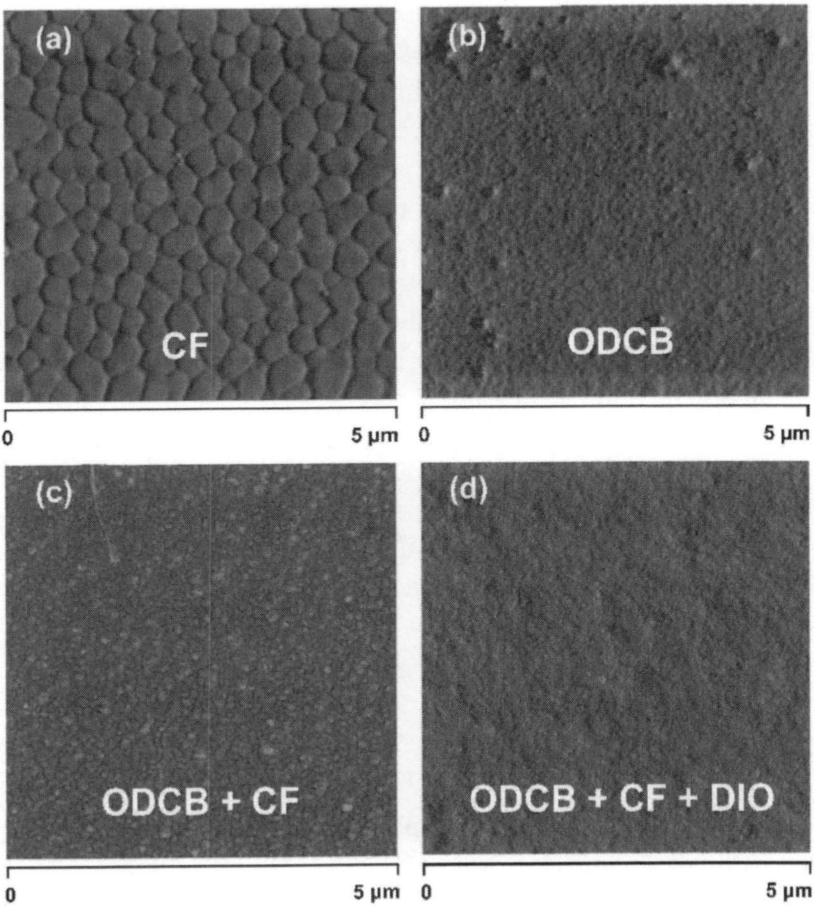

Figure 7. AFM phase image of PSTDPP/$PC_{71}BM$ blend film prepared by different solvents. (**a**) CF; (**b**) ODCB; (**c**) ODCB:CF (2:1); and (**d**) ODCB:CF (2:1) with 2.5% DIO. The size of the AFM image is 5 μm × 5 μm.

4. CONCLUSIONS

In summary, a new random copolymer containing two donors (SDT and BDT) and two acceptors (BT and DPP), named as PSTDPP, has been synthesized and fully characterized. As expected, the random polymer exhibits a broad absorption with a range of 500–800 nm and a proper

HOMO energy level, which ensure its applicability for fabrication of PSC devices. When using ODCB as processing solvent, a moderate PCE of 2.35% was obtained. To further optimize the device performances of PSTDPP/$PC_{71}BM$-based PSCs, different processing solvents (from ODCB to CF or ODCB/CF blend) were systematically investigated. It was found that the solvent mixtures of ODCB and CF with small amount of DIO as additive could offer a much better active layer with $PC_{71}BM$ enriched surface composition and improved morphology, which led to a significantly enhanced PCE of 4.38% compared with using ODCB as processing solvent. These results indicate that constructing random copolymer with multiple donor and acceptor monomers could provide a new general method for designing photovoltaic polymers with optimized optical and electrochemical properties. In addition, introducing proper mixed solvents to the active layer preparation is a very promising approach to improve the characteristics of the film, which would be very conducive to achieving high PSC performance.

ACKNOWLEDGMENTS

We acknowledge the financial support of the National Natural Science Foundation of China (Grant Nos. 21125420, 21474022 and 21101010) and the Ministry of Science and Technology of China (No. 2013 CB834700).

AUTHOR CONTRIBUTIONS

Kun Lu, Tianyi Zhao, Zhixiang Wei and Lei Jiang designed the experiments. Xiangwei Zhu, Benzheng Xia, Jin Fang and Yifan Zhao carried out the initial experiments. Xiangwei Zhu and Kun Lu carried out further experiments and wrote the manuscript. All authors discussed the data and commented on the manuscript.

CONFLICTS OF INTEREST

The authors declare no conflict of interest.

REFERENCES

1. Yu, G.; Gao, J.; Hummelen, J.C.; Wudl, F.; Heeger, A.J. Polymer photovoltaic cells: Enhanced efficiencies via a network of internal donor-acceptor heterojunctions. Science 1995, 270, 1789–1791.

2. Chen, J.; Cao, Y. Development of novel conjugated donor polymers for high-efficiency bulk-heterojunction photovoltaic devices. Acc. Chem. Res. 2009, 42, 1709–1718.

3. InganÄs, O.; Zhang, F.; Andersson, M.R. Alternating polyfluorenes collect solar light in polymer photovoltaics. Acc. Chem. Res. 2009, 42, 1731–1739.

4. Li, Y. Molecular design of photovoltaic materials for polymer solar cells: Toward suitable electronic energy levels and broad absorption. Acc. Chem. Res. 2012, 45, 723–733.

5. He, Z.; Xiao, B.; Liu, F.; Wu, H.; Yang, Y.; Xiao, S.; Wang, C.; Russell, T.P.; Cao, Y. Single-junction polymer solar cells with high efficiency and photovoltage. Nat. Photonics 2015, 9, 174–179.

6. Vohra, V.; Kawashima, K.; Kakara, T.; Koganezawa, T.; Osaka, I.; Takimiya, K.; Murata, H. Efficient inverted polymer solar cells employing favourable molecular orientation. Nat. Photonics 2015, 9, 403–408.

7. Dou, L.; Chen, C.-C.; Yoshimura, K.; Ohya, K.; Chang, W.-H.; Gao, J.; Liu, Y.; Richard, E.; Yang, Y. Synthesis of 5h-dithieno[3,2-b:2′,3′-d]pyran as an electron-rich building block for donor–acceptor type low-bandgap polymers.Macromolecules 2013, 46, 3384–3390.

8. Kroon, R.; Diaz de Zerio Mendaza, A.; Himmelberger, S.; Bergqvist, J.; Bäcke, O.; Faria, G.C.; Gao, F.; Obaid, A.; Zhuang, W.; Gedefaw, D.; et al. A new tetracyclic lactam building block for thick, broad-bandgap photovoltaics. J. Am. Chem. Soc. 2014, 136, 11578–11581.

9. Huang, Y.; Liu, F.; Guo, X.; Zhang, W.; Gu, Y.; Zhang, J.; Han, C.C.; Russell, T.P.; Hou, J. Manipulating backbone structure to enhance low band gap polymer photovoltaic performance. Adv. Energy Mater. 2013, 3, 930–937.

10. Mei, J.; Bao, Z. Side chain engineering in solution-processable conjugated polymers. Chem. Mater. 2014, 26, 604–615.

11. Price, S.C.; Stuart, A.C.; Yang, L.; Zhou, H.; You, W. Fluorine substituted conjugated polymer of medium band gap yields 7% efficiency in polymer–fullerene solar cells. J. Am. Chem. Soc. 2011, 133, 4625–4631.

12. Kang, T.E.; Kim, K.-H.; Kim, B.J. Design of terpolymers as electron donors for highly efficient polymer solar cells. J. Mater. Chem. A 2014, 2, 15252–15267.

13. Nielsen, C.B.; Ashraf, R.S.; Schroeder, B.C.; D'Angelo, P.; Watkins, S.E.; Song, K.; Anthopoulos, T.D.; McCulloch, I. Random benzotrithiophene-based donor-acceptor copolymers for efficient organic photovoltaic devices. Chem. Commun. 2012, 48, 5832–5834.

14. Jung, J.W.; Liu, F.; Russell, T.P.; Jo, W.H. Semi-crystalline random conjugated copolymers with panchromatic absorption for highly efficient polymer solar cells. Energ. Environ Sci. 2013, 6, 3301–3307.

15. Kang, T.E.; Cho, H.-H.; Kim, H.j.; Lee, W.; Kang, H.; Kim, B.J. Importance of optimal composition in random terpolymer-based polymer solar cells. Macromolecules 2013, 46, 6806–6813.

16. Wang, L.; Shi, S.; Ma, D.; Chen, S.; Gao, C.; Wang, M.; Shi, K.; Li, Y.; Li, X.; Wang, H. Improved photovoltaic properties of donor–acceptor copolymers by introducing quinoxalino[2,3-b′]porphyrin as a light-harvesting unit. Macromolecules2015, 48, 287–296.

17. Cho, N.; Yip, H.-L.; Jen, A.K.-Y. Morphology evolution by controlling solvent-solute interactions using a binary solvent in bulk heterojunction solar cells. Appl. Phys. Lett. 2013, 102, 233903.

18. Ye, L.; Zhang, S.; Ma, W.; Fan, B.; Guo, X.; Huang, Y.; Ade, H.; Hou, J. From binary to ternary solvent: Morphology fine-tuning of D/A blends in PDPP3T-based polymer solar cells. Adv. Mater. 2012, 24, 6335–6341.

19. Ye, L.; Jing, Y.; Guo, X.; Sun, H.; Zhang, S.; Zhang, M.; Huo, L.; Hou, J. Remove the residual additives toward enhanced efficiency with higher reproducibility in polymer solar cells. J. Phys. Chem. C 2013, 117, 14920–14928.

20. Son, S.K.; Lee, H.-S.; Ha, J.S.; Kim, K.H.; Son, H.J.; Ko, M.J.; Kim, H.; Lee, D.-K.; Kim, J.Y.; Lee, W.; et al. Effect of asymmetric solubility of diketopyrrolopyrrole-based polymers and PC71BMS in a binary solvent system on the performance of bulk heterojunction solar cells. Solar Energy Mater. Solar Cells 2014, 124, 232–240.

21. Wang, D.; Zhang, F.; Li, L.; Yu, J.; Wang, J.; An, Q.; Tang, W. Tuning nanoscale morphology using mixed solvents and solvent vapor treatment for high performance polymer solar cells. RSC Adv. 2014, 4, 48724–48733.

22. Van Franeker, J.J.; Turbiez, M.; Li, W.; Wienk, M.M.; Janssen, R.A.J. A real-time study of the benefits of co-solvents in polymer solar cell processing. Nat. Commun. 2015, 6.

23. Hou, J.; Chen, H.-Y.; Zhang, S.; Li, G.; Yang, Y. Synthesis, characterization, and photovoltaic properties of a low band gap polymer based on silole-containing polythiophenes and 2,1,3-benzothiadiazole. J. Am. Chem. Soc. 2008, 130, 16144–16145.

24. Hou, J.H.; Park, M.H.; Zhang, S.Q.; Yao, Y.; Chen, L.M.; Li, J.H.; Yang, Y. Bandgap and molecular energy level control of conjugated polymer photovoltaic materials based on benzo[1,2-b:4,5-b′]dithiophene. Macromolecules 2008, 41, 6012–6018.

25. Dhanabalan, A.; van Duren, J.K.J.; van Hal, P.A.; van Dongen, J.L.J.; Janssen, R.A.J. Synthesis and characterization of a low bandgap conjugated polymer for bulk heterojunction photovoltaic cells. Adv. Funct. Mater. 2001, 11, 255–262.

26. Wienk, M.M.; Turbiez, M.; Gilot, J.; Janssen, R.A.J. Narrow-bandgap diketo-pyrrolo-pyrrole polymer solar cells: The effect of processing on the performance. Adv. Mater. 2008, 20, 2556–2560.

27. Gupta, V.; Kyaw, A.K.K.; Wang, D.H.; Chand, S.; Bazan, G.C.; Heeger, A.J. Barium: An efficient cathode layer for bulk-heterojunction solar cells. Sci. Rep. 2013, 3, 1965.

28. Ye, L.; Zhang, S.; Zhao, W.; Yao, H.; Hou, J. Highly efficient 2D-conjugated benzodithiophene-based photovoltaic polymer with linear alkylthio side chain. Chem. Mater. 2014, 26, 3603–3605.

29. Liu, Y.; Zhao, J.; Li, Z.; Mu, C.; Ma, W.; Hu, H.; Jiang, K.; Lin, H.; Ade, H.; Yan, H. Aggregation and morphology control enables multiple cases of high-efficiency polymer solar cells. Nat. Commun. 2014, 5, 5293.

30. Choi, H.; Ko, S.-J.; Kim, T.; Morin, P.-O.; Walker, B.; Lee, B.H.; Leclerc, M.; Kim, J.Y.; Heeger, A.J. Small-bandgap polymer solar cells with unprecedented short-circuit current density and high fill factor. Adv. Mater. 2015, 27, 3318–3324.

31. Huo, L.; Hou, J.; Chen, H.-Y.; Zhang, S.; Jiang, Y.; Chen, T.L.; Yang, Y. Bandgap and molecular level control of the low-bandgap polymers based on 3,6-dithiophen-2-yl-2,5-dihydropyrrolo[3,4-c]pyrrole-1,4-dione toward highly efficient polymer solar cells. Macromolecules 2009, 42, 6564–6571.

32. Liang, Y.; Wu, Y.; Feng, D.; Tsai, S.-T.; Son, H.-J.; Li, G.; Yu, L. Development of new semiconducting polymers for high performance solar cells. J. Am. Chem. Soc. 2009, 131, 56–57.

33. Bürgi, L.; Turbiez, M.; Pfeiffer, R.; Bienewald, F.; Kirner, H.-J.; Winnewisser, C. High-mobility ambipolar near-infrared light-emitting polymer field-effect transistors. Adv. Mater. 2008, 20, 2217–2224.

34. Takacs, C.J.; Sun, Y.; Welch, G.C.; Perez, L.A.; Liu, X.; Wen, W.; Bazan, G.C.; Heeger, A.J. Solar cell efficiency, self-assembly, and dipole–dipole interactions of isomorphic narrow-band-gap molecules. J. Am. Chem. Soc. 2012, 134, 16597–16606.

35. Sun, Q.; Wang, H.; Yang, C.; Li, Y. Synthesis and electroluminescence of novel copolymers containing crown ether spacers. J. Mater. Chem. 2003, 13, 800–806.

36. Lu, K.; Fang, J.; Yu, Z.; Yan, H.; Zhu, X.; Zhang, Y.; He, C.; Wei, Z. Improving the performance of polymer solar cells by altering polymer side chains and optimizing film morphologies. Organic Electronics 2012, 13, 3234–3243.

37. He, Y.; Wu, W.; Zhao, G.; Liu, Y.; Li, Y. Poly(3,6-dihexyl-thieno[3,2-b]thiophene vinylene): Synthesis, field-effect transistors, and photovoltaic properties. Macromolecules 2008, 41, 9760–9766.

38. Lee, J.K.; Ma, W.L.; Brabec, C.J.; Yuen, J.; Moon, J.S.; Kim, J.Y.; Lee, K.; Bazan, G.C.; Heeger, A.J. Processing additives for improved efficiency from bulk heterojunction solar cells. J. Am. Chem. Soc. 2008, 130, 3619–3623.

39. Kiel, J.W.; Kirby, B.J.; Majkrzak, C.F.; Maranville, B.B.; Mackay, M.E. Nanoparticle concentration profile in polymer-based solar cells. Soft Matter. 2010, 6, 641–646.

40. Staniec, P.A.; Parnell, A.J.; Dunbar, A.D.F.; Yi, H.; Pearson, A.J.; Wang, T.; Hopkinson, P.E.; Kinane, C.; Dalgliesh, R.M.; Donald, A.M.; et al. The nanoscale morphology of a PCDTBT:PCBM photovoltaic blend. Adv. Energy Mater.2011, 1, 499–504.

41. Chung, H.-S.; Lee, W.-H.; Song, C.E.; Shin, Y.; Kim, J.; Lee, S.K.; Shin, W.S.; Moon, S.-J.; Kang, I.-N. Highly conjugated side-chain-substituted benzo[1,2-b:4,5-b′]dithiophene-based conjugated polymers for use in polymer solar cells.Macromolecules 2013, 47, 97–105.

42. Guo, X.; Cui, C.H.; Zhang, M.J.; Huo, L.J.; Huang, Y.; Hou, J.H.; Li, Y. High efficiency polymer solar cells based on poly(3-hexylthiophene)/indene-C70 bisadduct with solvent additive. Energ. Environ Sci. 2012, 5, 7943–7949.

4

SYNTHESIS AND CHARACTERISTICS OF NH_2-FUNCTIONALIZED POLYMER FILMS TO ALIGN AND IMMOBILIZE DNA MOLECULES

Hyung Jin Kim, In-Seob Bae, Sang-Jin Cho, Jin-Hyo Boo, Byung-Cheo Lee, Jinhee Heo, Ilsub Chung and Byungyou Hong

1School of Information and Communication Engineering, Sungkyunkwan University, Suwon, 440-746, Republic of Korea

2Metal Development Group, MDS Development Team, MDS Division, Samsung Techwin Co., LTD 42, Changwon, 642-716, Republic of Korea

3Department of Chemistry and Institute of Basic Science, Sungkyunkwan University, Suwon, 440-746, Republic of Korea

4Quantum Optics Lab, Korea Atomic Energy Research Institute, Daejeon, 305-353, Republic of Korea 5Korea Institute of Materials Science, Changwon, 641-831, Republic of Korea

ABSTRACT

We developed a method to use NH_2-functionalized polymer films to align and immobilize DNA molecules on a Si substrate. The plasma-polymerized cyclohexane film was deposited on the Si substrate according to the radio frequency plasma-enhanced chemical vapor deposition method using a single molecular precursor, and it was then treated by the dielectric barrier discharge method in a nitrogen environment under atmospheric pressure. Changes in the chemistry of the surface functional groups were studied using X-ray photoelectron spectroscopy and Fourier transformed infrared spectroscopy. The wettability of the surfaces was examined using dynamic contact angle measurements, and the surface morphology was evaluated using atomic force microscopy.

We utilized a tilting method to align λ-DNA molecules that were immobilized by the electrostatic interaction between the amine groups in NH_2-functionalized polymer films and the phosphate groups in the DNA. The DNA was treated with positively charged gold nanoparticles to make a conductive nanowire that uses the DNA as a template. We observed that the NH_2-functionalized polymer film was useful for aligning and immobilizing the DNA, and thus the DNA-templated nanowires.

KEYWORDS

DNA molecules, NH_2-functionalized polymer thin films, aniline-capped gold nanoparticles, gold nanowires.

1. INTRODUCTION

In the field of nanotechnology, DNA molecules are considered attractive building blocks for generating superstructures because the DNA itself is a nanowire with a nanoscale diameter of approximately 2 nm, and it has a very long linear structure with a well-defined polymeric sequence and many functional groups [1, 2]. Therefore, the metal nanowires that are fabricated by the conjugation of DNA molecules and nanosize metal nanoparticles have been extensively investigated for their application to the highly ordered electronic components of nanocircuitry and/or nanodevices [3–7]. The technique of aligning and immobilizing the DNA molecules on various substrates is an important technique; however, this technique is very difficult to control. The technique of controlling and manipulating DNA molecules with nanometer resolution is a subject of priority in the field of DNA-based nanotechnology; many research groups have studied the application of the latter technique in nanotechnology [8]. Recently, it became possible to align and immobilize DNAs by various physical methods such as electric or dielectric force [9], microcontact printing (μCP) [10, 11], fluid flow [12], and molecular combing/surface tension [13, 14]. Surface modification, such as silanization, has also been used to assist the immobilization and the alignment of DNA molecules [14–17]. However, self-assembled monolayers that are formed by the silanization method have unstable physical and chemical conditions because they depend on conditions such as temperature and humidity during coating.

In this paper, we present a new procedure to reproducibly align and immobilize DNAs using NH_2-functionalized polymer films that are deposited on a Si substrate according to the radio frequency [RF] plasma-enhanced chemical vapor deposition [PECVD] method using cyclohexane, and it is then treated by the dielectric barrier discharge [DBD] method in a nitrogen [N_2] environment. The key point of this technique is the alignment of the DNA through the interactions between the phosphate groups in the DNA and the amine groups on the surface of the NH_2-functionalized polymer film. It is observed on how well the DNA is arranged on the polymer film's surface by atomic force microscopy [AFM] examination; moreover, we discuss the feasibility of this technique for nanowire synthesis for its application to nanodevices.

2. EXPERIMENTAL DETAILS

A solution of DNA (16 μm long - 48,502 bp; Bio Basic Inc., Markham, Ontario, Canada) was prepared in a concentration of 10 ng/μL with TE buffer (10 mM tris-HCl and 1 mM EDTA, pH 8.0). Aniline (Sigma-Aldrich Corporation, St. Louis, MO, USA)-capped Au nanoparticles [AN-AuNPs] were prepared based on the conventional reduction of $HAuCl_4$ (Sigma-Aldrich Corporation, St. Louis, MO, USA) using aniline as a reducer [14, 17, 18]. The plasma-polymerized cyclohexane films, with an average thickness of 200 nm, were deposited on the Si substrate at a RF power of 20 to approximately 50 W according to the PECVD method using cyclohexane, and these films were then treated by the DBD method in a N_2 environment under atmospheric pressure at a RF power of 150 W for 60 s to form the amine groups on the polymer film surface. The NH_2-functionalized polymer surface plays an important role in attaching the DNA onto the substrate through a strong electrostatic interaction between the amine groups of the sample surface and the negatively charged phosphate backbone of the DNA [18]. Changes in the chemistry of the surface functional groups were studied using X-ray photoelectron spectroscopy [XPS] and Fourier transform infrared spectroscopy [FT-IR] spectroscopy. The wettability of the surfaces was examined using dynamic contact angle measurements, and the surface morphology was evaluated using AFM (SPA 400, SII Nanotechnology Inc., Sunto-gun, Shizuoka, Japan).

DNA alignment was carried out according to the tilting method, and this resulted in highly aligned DNA patterns on the substrate [16, 18]. A λ-DNA solution of 10 μL was dropped on the surface-treated polymer film, and the sample was then tilted to almost 90°. DNA molecules were stretched and aligned along the direction of the sample's tilting. For the synthesis of the conductive nanowire, DNA molecules were treated with AN-AuNPs of 60 μL for 30 min; AN-AuNPs of 60 μL were prepared based on the conventional reduction of $HAuCl_4$ using aniline as a reducer [14, 17, 19]. Next, the sample is rinsed with double distilled water. During this treatment, the DNA-templated gold nanowires [AuNWs] are fabricated through the electrostatic interaction between the positively charged amine groups of the AuNPs and the negatively charged phosphate groups in the DNA. The observation by AFM showed that the DNAs stretched on the NH_2-functionalized polymer films and that the AuNPs were assembled along DNA molecules.

3. RESULTS AND DISCUSSIONS

The purpose of this work was to confirm the possibility of aligning and immobilizing DNA molecules on the NH_2-functionalized polymer thin films formed by the conventional PECVD and DBD methods; this was motivated by the possibility that the negatively charged phosphate backbone of the DNA would electostatically interact with the positively charged polymer film with an amine group.

XPS spectra in wide scan and high resolution were recorded for the plasma-untreated and plasma-treated polymer thin films, which were deposited on the Si substrate at a RF power of 30 W by the PECVD method using cyclohexane, and then, they were treated by DBD method in N_2 at a RF power of 150 W for 60 s to form the functional groups on the surface region and to show the subsequent changes of the chemistry on the surface that were introduced by the N_2 plasma treatment. Figure 1a mainly shows the signals in XPS that correspond to O1s (532 eV), C1s (285 eV), and N1s (400 eV). The inset in Figure 1a shows enlarged N1s XPS spectra for the plasma-treated and untreated polymer films. As seen in Figure 1a, there were no N1s signals on the plasma-untreated polymer films. However, N_2 plasma-treated polymer films showed N1s signals and strong reduction of the C1s signal. The N1s peak for the N_2 plasma-treated polymer film was resolved into three peaks at 399.2, 400.4, and 401.7 eV (Figure 1b). These peaks correspond to N-C, N = C, and N-C = O with relative compositions of 10.5%, 44.6%, and 44.9%, respectively [18,20]. This indicates the presence of nitrogen atoms with at least three different binding energies.

Details of different carbon functionalities at the surface were also determined from the C1s high-resolution XPS spectrum in Figure 1c, d. The C1s spectra for the untreated and treated polymer films were deconvoluted into five component peaks, which were C = C at 284.6 eV, C-C at 285 eV, C-O at 286.2 eV (includes a small C-N bond contribution at 285.9 eV), C = O at 287.7 eV (includes a small C = N bond contribution at 288.0 eV), and O-C = O at 289.4 eV [18–21].

Due to the N_2 plasma treatment, there was a significant incorporation of nitrogen-containing groups. The O1s signal was unaffected; however, it was likely that the incorporation of oxygen moieties occurred upon exposure to air before the plasma treatment, but at the expense of oxygenated molecules at the surface lost during plasma exposure. Consequently, the intensity of the C1s spectrum was affected such that the intensities of C = O and O-C = O bonds decreased. This was a concomitant increase of the peak intensity at 288.6 eV corresponding to the O = C-N bond. It indicates significant incorporation of nitrogen-containing groups, but the exact identification of functionalities that are bound to the surface is difficult. In addition to C-O type moieties, the component at 286.6 eV may include C-NH_2, C-NH, or C-N = C groups. The N-C = O peak located at 288.6 eV could imply N-C-O and $CONH_2$ type functionalities. The results were likely that nitrogen-containing species have been formed in the N_2plasma-treated polymer film (Figure 1).

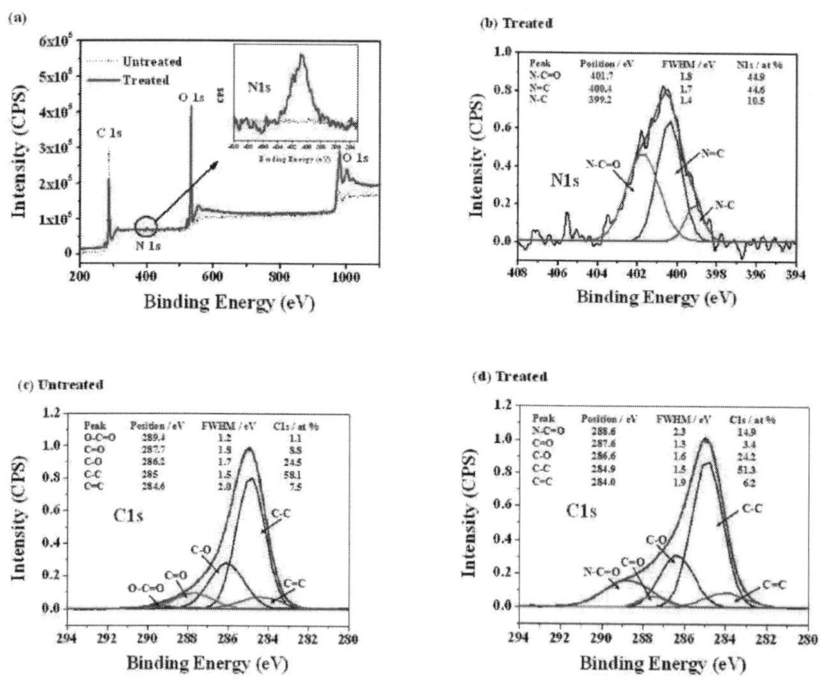

Figure 1 XPS spectra. (**a**) A wide scan XPS spectra of the N_2 the plasma-untreated and treated polymer films. The inset shows enlarged N1s XPS spectra in Figure 1. (**b**) High-resolution XPS spectra of N1s peaks for the plasma-treated polymer film. (**c**) High-resolution XPS spectra of C1s peaks for the plasma-untreated polymer film. (**d**) High-resolution XPS spectra of C1s peaks for the N_2 plasma-treated polymer film.

Figure 2 shows the attenuated total reflection method of the FT-IR absorption spectra over the range of 600 to approximately 4,000 cm^{-1}

on the plasma-untreated and plasma-treated polymer films according to the DBD plasma treatment method. The spectra exhibit an absorption peak for increasing the N-H stretching vibration at 3,400 cm^{-1} and the N-H bending vibration at 1,650 cm^{-1}, and the spectra exhibit an absorption peak for decreasing the C-H stretching vibration at 2,900 cm^{-1} and the C-H bending vibration at 1,400 cm^{-1} in the case of the N_2 plasma-treated polymer film, as shown in Figure 2a. The results further prove that the nitrogen atoms were covalently grafted to the polymer main chains to form new amine groups (NH_2), which correspond to the XPS results, because N ions are chemically grafted onto the polymeric structure and are highly reactive with H ions.

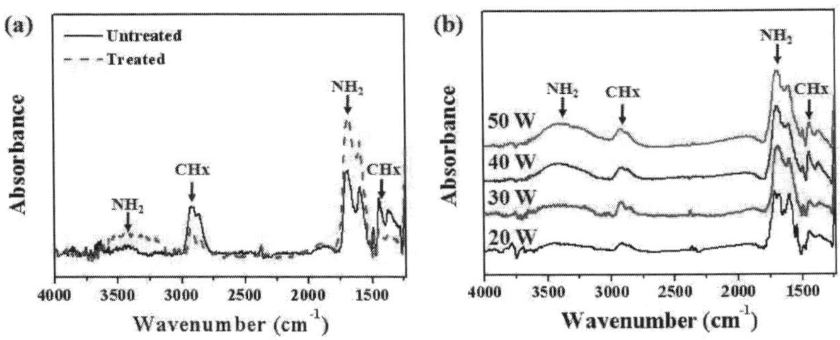

Figure 2 FT-IR spectra. (**a**) FT-IR spectra of the N_2 plasma-untreated and treated polymer films at a RF power of 150 W for 60 s. (**b**) FT-IR spectra for N_2 plasma-treated polymer films that were synthesized with the varied RF power (20 to approximately 50 W).

The average surface roughness of the NH_2-functionalized polymer films, which were deposited on the Si substrate at a RF power of 30 W by PECVD method using cyclohexane and then treated by the DBD method in N_2 at a RF power of 150 W for 60 s, was 0.45 ± 0.03 nm according to the AFM measurements. This smooth surface simplified the AFM measurements of the well-aligned DNA on the Si wafer. Tilting the sample induced an air-water interface motion, and thus, the DNA molecules were aligned along the motion direction of the droplet. The average height of the aligned DNA molecules was 0.41 ± 0.17 nm (Figure 3b), and this value was consistent with the height of a double-stranded

DNA from the AFM measurements that were reported by other groups [10, 11]. λ-DNAs were confirmed to be aligned on the NH_2-functionalized polymer films as expected, and they were, more or less, evenly distributed.

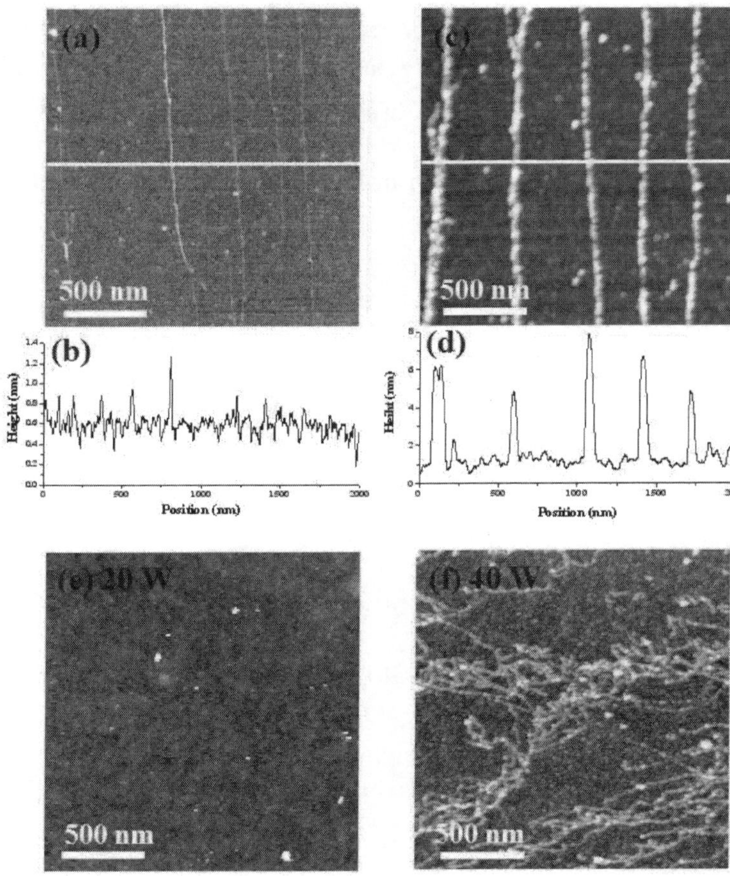

Figure 3 AFM images. (**a**) AFM images of the DNA molecules that were stretched and immobilized on the N_2 plasma-treated polymer film synthesized with 30 W of RF power. (**b**) The scan profile along the white line of the image (a). (**c**) AFM image of AuNWs formed along the stretched DNA molecules on the N_2 plasma-treated polymer film that was synthesized with 30 W of RF power. (**d**) The scan profile along the white line of image (c). (**e, f**) AFM images showing the dependence of DNA stretch and alignment on NH_2-functionalized polymer films that were synthesized (e) at 20 W power and (f) at 40 W power, respectively, and then treated with DBD at a RF power of 150 W for 60 s.

In addition to the alignment and immobilization of DNA molecules on the NH_2-functionalized polymer film, another notable point proposed in this work is that the conductive AuNWs can be formed by the treatment of DNAs with the aniline [AN]-capped AuNPs. DNAs aligned on the NH_2-functionalized polymer film were treated by a solution of AuNPs (5 nm) whose surface was positively charged after they were treated with AN solution, and thus, AuNPs were strongly attached to the negatively charged DNAs. However, they were weakly bound on the NH_2-functionalized polymer film and easily washed away by rinsing. Figure 3c, d shows AFM images of AuNWs aligned along the well-stretched DNAs on the NH_2-functionalized polymer film. The average height of the DNA-templated AuNWs was 5.3 ± 0.5 nm. This height was similar to the average diameter of aniline-capped AuNPs, which were observed with the TEM, used in this experiment.

Figure 2b shows FT-IR spectra of the N_2 plasma-treated polymer films that were synthesized by varying the RF power (20 to approximately 50 W) using the PECVD method and were treated in N_2 plasma using the DBD method at a RF power of 150 W for 60 s. The spectra exhibit that the absorption peaks of N-H (at 3,400 cm^{-1} and 1,650 cm^{-1}) depended on plasma power, which was a parameter for polymer film synthesis, and those absorption peaks increased with an increase in the plasma power. It was reported that using high plasma power in polymer film synthesis has caused higher degrees of C = C bond contents due to high cross-linking between radicals of monomer molecule compared to using low plasma power [22]. Hence, it was expected that C = C bond, which has a π-bonding, can easily be broken by N_2 plasma. This process can generate the cation radical at the surface and then more amine groups on surface of the plasma-polymerized film during plasma treatment using DBD process in N_2environment than the low plasma power.

Several types of experiments were carried out to prove that plasma power affects the alignment and immobilization of the DNA and the surface energy for the NH_2-functionalized polymer films. First, the average coverage of the aligned DNA on the surface depends on the RF power. In a low power range (≤20 W), the negatively charged DNA molecules were weakly bound to the NH_2-functionalized polymer thin film, and they were easily swept off from the surface during the alignment process (Figure 3e). However, a high power (≥40 W) caused the DNA molecules to be formed in bundles and random networks, and thus, they were not well stretched (Figure 3f). It was reported that the surface tension and the density of amine groups on the surface determined the degree of straightness and density of the aligned DNA molecules [10, 12, 23]. Therefore, it was considered that the proper surface treatment of the polymer film determined this aspect, and thus, the appropriate interaction between the surface of the NH_2-functionalized polymer thin film and the DNA molecules would give excellent alignment of DNA molecules.

In addition, the surface energy was able to be controlled by RF power. The contact angle is one of the surface's sensitive properties that detects any changes at the modified surface. It is well known that the water contact angle increases as the surface energy is reduced. From the contact angle and AFM results, it was observed that the contact angle for the treated films decreased from 64° to 41°, but the root mean square [rms] roughness increased from 0.38 nm to 0.54 nm as the deposition power increased (Figure 4). This indicates that the N_2 plasma-treated polymer films had higher surface energy with an increase in the deposition power. It was also demonstrated that such treatment and high plasma power were able to improve the ability of polymer films to be bound with other materials. Our results prove that RF power affects the density of the amine groups that are formed on the polymer thin films, and they play an important role in aligning and immobilizing the DNA on the NH_2-functionalized polymer films.

4. CONCLUSIONS

For the first time, we have proposed to use the NH_2-functionalized polymer film for the alignment and the immobilization of DNA molecules for the purpose of applying them to nanodevices. The polymer film was deposited on a Si substrate by the PECVD method using a single molecular precursor, and the film was then treated using the DBD plasma surface treatment (or modification) method in N_2 plasma under atmospheric pressure to form the amine groups on the surface. XPS and FT-IR spectra showed that the technique has been successful in forming nitrogen-containing groups by the N_2 plasma, and this wasn't seen in the untreated material. The treated polymer surface was smooth enough (roughness was 0.45 ± 0.17 nm) to allow proper AFM measurements of the well-aligned DNA on the sample wafer. The DNA was well aligned and immobilized on the surface of the NH_2-functionalized polymer films according to the tilting method. It was demonstrated that the N_2 plasma treatment has been successful in forming the NH_2-functionalized polymer films. In addition, the plasma power that was used to synthesize the polymer film played an important role in immobilizing the DNA on the NH_2-functionalized polymer films as intended.

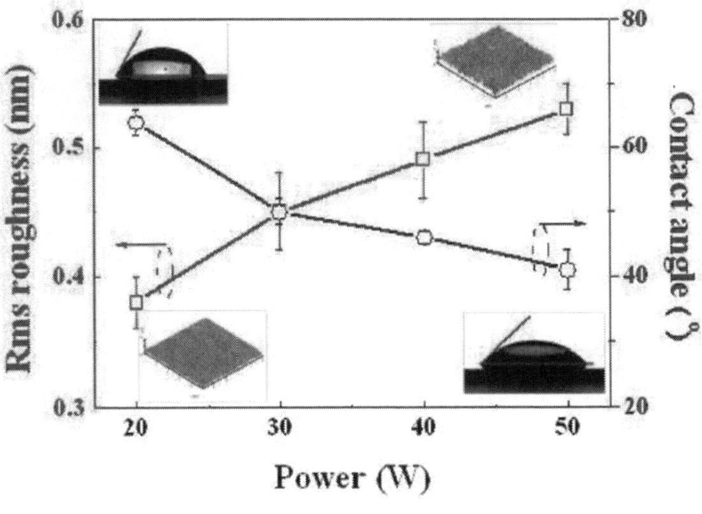

Figure 4 Variation of rms roughness and contact angle for NH_2-functionalized polymer film with plasma RF power.

Therefore, we believe that the NH_2-functionalized polymer film could be a promising material for organizing the DNA-based building blocks into structures that are needed for wiring and interconnecting functional nanodevices and/or biosensors. In addition, the fabrication technology could be easily adapted for arraying nanowires into more complex crossed structures and for making nanowires of other materials ordered.

ACKNOWLEDGEMENTS

This work was supported by the Korea Science and Engineering Foundation (KOSEF) grant funded by the Korean government (MEST) (R01-2008-000-10690-0), and also by the Nuclear R&D program through the Korean Science and Engineering Foundation funded by the Ministry of Science & Technology (Project No. 2007-01582).

COMPETING INTERESTS

The authors declare that they have no competing interests.

AUTHORS' CONTRIBUTIONS

HJK and BH conceived the study. ISB, SJC, and JHB carried out the experiments. JH and IC contributed to the analysis study. BCL drafted the manuscript. All authors are involved in revising the manuscript and approved the final version.

REFERENCES

1. Nakao H, Hayashi H, Iwata F, Karasawa H, Hirano K, Sugiyama S, Ohtani T: Fabricating and aligning π-conjugated polymer-functionalized DNA nanowires: atomic force microscopic and scanning near-field optical microscopic studies. *Langmuir* 2005, 21: 7945. 10.1021/la050145p

2. Braun E, Eichen Y, Sivan U, Ben-Uoseph G: DNA-templated assembly and electrode attachment of a conducting silver wire. *Nature* 1998, 391: 775.

3. Richter J, Mertig M, Pompe W, Monch I, Schackert HK: Construction of highly conductive nanowires on a DNA template. *Appl Phys Lett*2001, 78: 536.

4. Harnack O, Ford WE, Yasuda A, Wessels JM: Tris(hydroxymethyl) phosphine-capped gold particles templated by DNA as nanowire precursors. *Nano Lett* 2002, 2: 919.

5. Keren K, Krueger M, Gilad R, Ben-Yoseph G, Sivan U, Braun E: Sequence-specific molecular lithography on single DNA molecules. *Science* 2002, 297: 72.

6. Porath D, Bezryadin A, de Vries S, Dekker C: Direct measurement of electrical transport through DNA molecules. *Nature* 2000, 403: 635.

7. Nickels P, Dittmer WU, Beyer S, Kotthaus JP, Simmel FC: Polyaniline nanowire synthesis template by DNA. *Nanotechnology* 2004, 15: 1524.

8. Nakao H, Hayashi H, Yoshino T, Sugiyama S, Otobe K, Ohtani T: Development of novel polymer-coated substrates for straightening and fixing DNA. *Nano Lett* 2002, 2: 475.

9. Washizu M, Kurosawa O: Electrostatic manipulation of DNA in microfabricated structures. *IEEE Trans Ind Appl* 1990, 26: 1165.

10. Zhang J, Ma Y, Stachura S, He H: Assembly of highly aligned DNA strands onto Si chips. *Langmuir* 2005, 21: 4180.

11. Nakao H, Gad M, Sugiyama S, Otobe K, Ohtani T: Transfer-printing of highly aligned DNA nanowires. *J Am Chem Soc* 2003, 125: 7162.

12. Deng Z, Mao C: DNA-templated fabrication of 1D parallel and 2D crossed metallic nanowire arrays. *Nano Lett* 2003, 3: 1545.

13. Michalet X, Ekong R, Fougerousse F, Rousseaux S, Schurra C, Hornigold N, Slegtenhorst M, Wolfe J, Povey S, Beckmann JS, Bensimon A: Dynamic molecular combing: stretching the whole human genome for high-resolution studies. *Science* 1997, 277: 1518.

14. Shin M, Kwon C, Kim SK, Kim HJ, Roh Y, Hong B, Park JB, Lee H: Formation of λ -DNA's in parallel- and crossed-line arrays by molecular combing and scanning-probe lithography. *Nano Lett* 2006, 6: 1334.

15. Kim HJ, Roh H, Hong B: Controlled gold nanoparticle assembly on DNA molecule as template for nanowire formation. *J Vac Sci Technol A*2006, 24: 327.

16. Heim T, Mélin T, Deresmes D, Vuillaume D: Localization and delocalization of charges injected in DNA. *Appl Phys Lett* 2004, 85: 2637.

17. Nakao H, Shiigi H, Yamamoto Y, Tokonami S, Nagaoka T, Sugiyama S, Ohtani T: Highly ordered assemblies of Au nanoparticles organized on DNA. *Nano Lett* 2003, 3: 1391.

18. Wilsona DJ, Rhodesb NP, Williamsa RL: Surface modification of a segmented polyetherurethane using a low-powered gas plasma and its influence on the activation of the coagulation system. *Biomaterials* 2003, 24: 5069.

19. Kim HJ, Roh Y, Hong B: Fabrication and characterization of DNA-templated conductive gold nanoparticle chains. *J Appl Phys* 2009, 105: 074302.

20. Weibel DE, Vilani b C, Habert AC, Achete CA: Surface modification of polyurethane membranes using RF-plasma treatment with polymerizable and non-polymerizable gases. *Surf Coat Tech* 2006, 201: 4190.

21. Yang D-Q, Poulin S, Martinu L: Microscale chemical and electrostatic surface patterning of Dow Cyclotene by N_2plasma. *Appl Surf Sci*2005, 242: 419.

22. Bae I-S, Cho S-H, Lee S-B, Kim Y, Boo J-H: Growth of plasma-polymerized thin films by PECVD method and study on their surface and optical characteristics. *Surf Coat Technol* 2005, 193: 142.

23. Abramchuk SS, Khokhlov AR, Iwataki T, Oana H, Yoshikawa K: Direct observation of DNA molecules in a convection flow of a drying droplet. *Europhysics Lett* 2001, 55: 294.

5

ELECTROSPRAY DEPOSITION OF POLYMER THIN FILMS FOR ORGANIC LIGHT-EMITTING DIODES

Wontae Hwang, Guoqing Xin, Minjun Cho, Sung Min Cho and Heeyeop Chae

1School of Chemical Engineering, Sungkyunkwan University (SKKU), Suwon, 440-746, South Korea

2SKKU Advanced Institute of Nanotechnology (SAINT), Sungkyunkwan University (SKKU), Suwon, 440-746, South Korea

ABSTRACT

Electrospray process was developed for organic layer deposition onto polymer organic light-emitting diode [PLED] devices in this work. An electrospray can be used to produce nanometer-scale thin films by electric repulsion of microscale fine droplets. PLED devices made by an electrospray process were compared with spin-coated ones. The PLED device fabricated by the electrospray process showed maximum current efficiency of 24 cd/A, which was comparable with that of the spin-coating process. The electrospray process required a higher concentration of hole and electron transport materials in the inks than spin-coating processes to achieve PLED maximum performance. Photoluminescence [PL] at 407 nm was observed using electrosprayed poly(*N*-vinyl carbazole) films, whereas a peak at 410 nm was observed with the spin-coated ones. Similar difference in peak position was observed between aromatic and nonaromatic solvents in the spin-coating process. PLED devices made by the electrospray process showed lower current density than that of spin-coated ones. The PL peak shift and reduced current of electrosprayed films can therefore be attributed to the conformation of the polymer.

KEYWORDS

organic light-emitting diodes, electrospray, polymer, conformation.

1. INTRODUCTION

During the last two decades, intense academic and industrial research has been devoted on organic light-emitting devices [OLEDs] due to their potential applicability to flat panel displays and solid-state lighting [1–3]. Small areas of less than 5-in. devices have been commercialized recently. Currently, all commercial processes adopt vacuum evaporation processes for both organic and metal layer formations. However, vacuum evaporation processes are significantly limited in large-area processing as well as hardware cost and require a considerable material. These limitations and drawbacks prevent OLEDs from being applicable to large-area device fabrication. Much effort has been made for solution process development that can form nanoscale-thin organic films with a large area while minimizing material wastes. A laboratory-scale spin-coating process is one of the most well known and widely used solution processes. Various solution printing techniques have been developed such as ink-jet printing, nozzle printing, screen printing, gravure printing, and so on. Thin-film transfer process, organic vapor deposition, and blade coating are also being developed as large-area, cost-effective alternatives [4–10]. However, challenges remain relating to good uniformity over a large-area and multilayer formation without buffer or cross-linking materials for commercial development of organic displays and lighting devices.

Recently, electrospray process has gained much attention as a solution process for organic and inorganic thin films, and a few research groups have reported applications to organic device fabrications [11–13]. In the electrospray process, a liquid flow is injected into the nozzle with an electric field applied between the nozzle tip and ground plate, and microscale monodisperse fine droplets are generated due to repulsion forces between like charges in the drops. The size of droplets can be controlled by adjusting the flow rate and electric field applied to the injection nozzles and substrates, and the diameter of the droplets can be as small as several hundred nanometers in scale [14–16]. The electrospray process with vapor treatment has been applied to organic photovoltaic fabrication, while comparable power conversion efficiency has been reported with the spin-coating process [13]. Additionally, applicability of the electrospray process to organic thin films in OLEDs was demonstrated in small-scale devices [12].

In this work, we demonstrated nanoscale-thick organic thin films using the electrospray process as a solution process alternative. Polymer LEDs [PLEDs] were fabricated and compared with those made by the spin-coating process. The electrospray process can be considered as an effective process for patterning, multilayer stacking, and continuous processing of organic thin films.

2. EXPERIMENTAL DETAILS

For PLED fabrication, we used a blended solution of poly(*N*-vinyl carbazole) [PVK], 2-(4-biphenylyl)-5-(4-tert-butylphenyl)-1, 3, 4-oxadiazole [PBD], *N*, *N*'diphenyl-*N*, *N*'-Bis(3-methylphenyl)-[1, 1-biphenyl]-4, 4'-diamine [TPD], and tris(2-(4-tolyl) phenylpyridine) iridium [$Ir(mppy)_3$] dissolved in chlorobenzene [CB], dichlorobenzene [DCB], or a mixture of CB and 1, 2-DCB. Two different types of inks were formulated as shown in Table 1. PLED devices were fabricated with a ratio of PVK/PBD/TPD/$Ir(mppy)_3$ = 61:24:9:6 (ink 1) for both the spin-coating and electrospray processes. The PBD/TPD ratio was increased for the electrospray process. CB is best for the spin-coating process among the solvents mentioned earlier, and DCB addition is required for the electrospray process.

Table 1 Ratio of the PLED ink

	Ink 1[a]	Ink 2[a]
PVK	61	41.5
PBD	24	41.5
TPD	9	14.8
$Ir(mppy)_3$	6	4.2

[a]Each figure means weight percent of materials in the solution, and the main difference between Ink 1 and Ink 2 is the ratio of hole and electron transport materials. PVK, poly(*N*-vinyl carbazole); PBD, 2-(4-biphenylyl)-5-(4-tert-butylphenyl)-1, 3, 4-oxadiazole; TPD, *N, N'*diphenyl-*N, N'*-Bis(3-methylphenyl)-[1, 1-biphenyl]-4, 4'-diamine; $Ir(mppy)_3$, tris(2-(4-tolyl) phenylpyridine) iridium.

The layer structure of the devices was as follows: indium tin oxide [ITO]/ poly(3, 4-ethylenedioxythiophene) poly(styrenesulfonate) [PEDOT:PSS]/active layer/metal cathode. A hole-injection PEDOT:PSS layer of 30-nm thick was first spin-coated on pre-cleaned ITO substrates and then baked at 120°C for 20 min. About 80-nm-thick emissive layer [EML] was formed by electrospray or spin-coating process and then annealed at 80°C for 30 min. A schematic diagram of the experimental setup for the electrospray process is shown in Figure 1. For the electrospray process, the blend ink was injected through the nozzle at a rate of 30 μl/min, and about 3 kV was applied to break the meniscus formed at the tip of the nozzle. The distance between the tip of the nozzle (150 μm in diameter) and the substrate was maintained at 3 to 4 cm. During spraying, the nozzle and substrate were fixed, and the thickness of EML was controlled by varying the deposition time. For the comparative spin-coating process, the solution was spin-coated at 2, 000 rpm for 20 s. All the experiments were carried out at 20°C to 25°C and a humidity of 30% to 35%. The interlayer and cathode were thermally evaporated on the top of the EML at a pressure of 2×10^{-5} Torr.

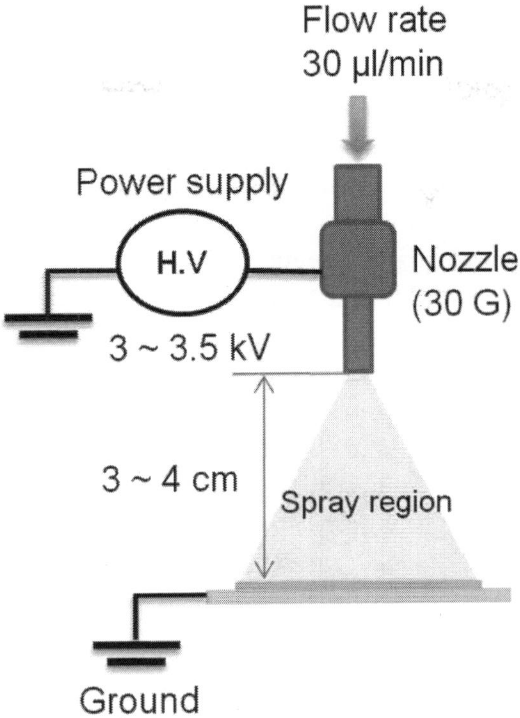

Figure 1 Schematic description of the electrospray process used for PLED fabrication.

All electrical measurements were performed under ambient conditions. Device performance was measured using a source measure unit (2400, Keithley Instruments, Inc., Cleveland, OH, USA) and a luminance meter (CS100, Konica Minolta Sensing, Inc., Sakai, Osaka, Japan). Photoluminescence [PL] spectra were collected with a monochromatized 150-W Xe light source (FP-6200, Jasco International Co. Ltd., Hachioji, Tokyo, Japan), and a wavelength of 335 nm was used for analysis of PL excitation. The thickness and roughness of PVK films were determined by a surface profiler (Alpha-Step, KLA-Tencor Corporation, Milpitas, CA, USA) and by atomic force microscopy (Digital Instruments, Santa Barbara, CA, USA), respectively.

3. RESULTS AND DISCUSSION

In electrospray deposition, it is critical to choose proper solvents to make fine droplets and uniform thin films. Evaporation rate of the solvent, which is determined by its boiling point, is an important factor to consider. Dielectric constant is another crucial physical property of solvents in determining the size of droplets and droplet dispersion. Further, high dielectric constant and high boiling point are preferred in electrospray processing. Specifically, 1, 2-DCB (dielectric constant [ε] = 10, boiling point = 180°C) is added to CB (ε = 5, boiling point = 130°C) [17] to balance solubility and droplet control. The concentration of droplets at the time of deposition is then controlled by managing the droplet size. Droplets containing little solvent or almost dry particles can be deposited when a low-boiling point solvent is used at a slow flow rate. This mode of operation is called the dry mode, and a few groups have reported a dry-mode operation [18]. In this work, droplets containing a rather large fraction of solvents were transferred to substrates, and therefore, the films required time for solvent evaporation after the deposition. As shown in Figure 2, 70- to 150-nm-thick light-emitting organic layers with about 1-nm surface roughness were formed by the electrospray process in this work.

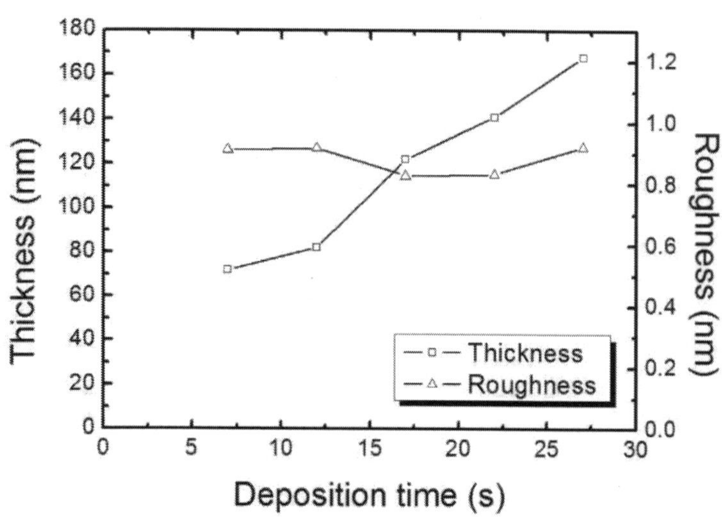

Figure 2 PVK film thickness and roughness as a function of deposition time.

To investigate the characteristics of organic light-emitting layers made by the electrospray deposition process, we adopted organic light-emitting materials of PVK as a host polymer with $Ir(mppy)_3$ added as a dopant. Further, TPD and PBD were added to facilitate a hole and electron transport. We performed experiments with our standard composition, ink (1) in Table 1, and the device made with the spin-coating process with ink (1) is shown with sample 1 in Figure 3. When we used the same ink (1) in the electrospray process, the current efficiency significantly decreased by up to 38% (sample 2). We believe from the cause was an imbalance in charge transport in the device, and we were able to achieve similar performance by formulating ink (2) with increased hole and electron transport materials as shown with sample 5 in Figure 3. Samples 5 and 6 show the effects of the solvent in the electrospray process. Better device performance was achieved with a solvent mixture of CB/DCB = 5:3. DCB helps maintain a stable and uniform spray, but too much DCB lowers device performance since it cannot be removed sufficiently during the annealing process due to its high boiling point. This result indicates that solvent optimization can regulate film and device performance made by the electrospray process. The summary of device performances is in Table 2.

To investigate the difference between electrospray and spin-coating processes, PVK thin films were formed by two different processes, and their electric and PL characteristics were investigated. As mentioned earlier, development of a proper solvent is an important step in the electrospray process. Accordingly, we also investigated the effects of solvents on electric and PL film characteristics. Figure 4 shows the effects of solvents and process method on PL characteristics. We chose two different types of solvents: an aromatic solvent which comprised a mixture of CB and DCB and non-aromatic solvents, chloroform [CF] and 1, 2-dichloroethane [DCE]. The mixture of CB and DCB was selected as it was optimum for our device structure and process. The PL peaks of the thin films were observed at 407 nm with CF and DCE and at 410 nm with the aromatic solvent mixture. Using the same aromatic solvent, PVK films processed by the electrospray process showed PL peaks at 407 nm, while those made by spin-coating showed PL peaks at 410 nm. Qian et al. [19] reported a peak shift of PVK due to polymer conformation caused by the solvent effect. We believe that a similar polymer conformation effect results not only from the solvent, but also from the processing method as shown in this work. Figure 5 shows the current-voltage characteristics of the PVK film using different solvents and processing methods. At the same thickness, the thin films using the aromatic solvent (mixture of CB and DCB) showed a higher current density than those processed with the DCE nonaromatic solvent. In terms of processing methods, the PVK film processed with the electrospray method showed a lower current density than that of spin-coated ones. These results suggest that polymer conformation of PVK thin films was affected by the processing method as well as by the solvent's molecular structure as also indicated in PL results in Figure 4. Based on these results, we speculate that the properties of thin films made by the electrospray process are similar to those of spin-coated films made with DCE.

4. CONCLUSION

We have demonstrated polymer organic LEDs using an electrospray process as a solution process alternative. Device performance comparable to spin-coating was achieved with the electrospray process, which can be considered as a scalable large-area process alternative. However, the electrospray process requires elaborate choices involving solvents and higher concentrations of hole and electron transport materials in active materials. A high dielectric constant and a high-boiling point solvent like DCB are preferred in electrospray processing. Accordingly, the device based on the electrospray process had better performance when we used the ink having higher concentrations of hole and electron transport materials. The electrospray process can be considered as a viable solution for large-area organic thin-film formation technology in the future.

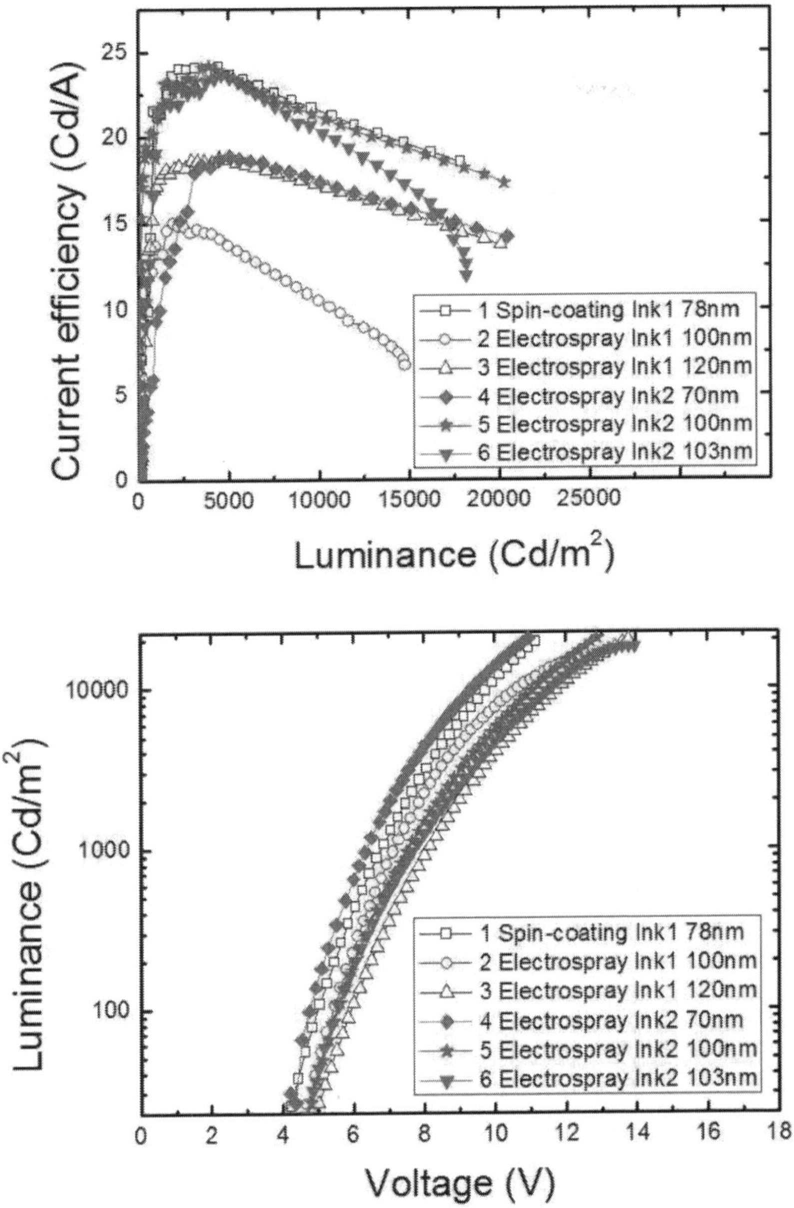

Figure 3 Device performance of the devices made by spin-coating or electrospraying. The thicknesses of EML were as follows: (1) 78, (2) 100, (3) 120, (4) 70, (5) 100, and (6) 103 nm.

Table 2 Summary of device performances

Device	Process	Ink	Solvent	EML (nm)	Cd/A (max)	Lm/W (max)
1	Spin-coating	Ink 1	CB	78	24	10
2	Electrospray	Ink 1	CB/DCB (5:3)	100	15	6
3	Electrospray	Ink 1	CB/DCB (5:3)	120	18	6.6
4	Electrospray	Ink 2	CB/DCB (5:3)	70	18	7.5
5	Electrospray	Ink 2	CB/DCB (5:3)	100	24	9.2
6	Electrospray	Ink 2	CB/DCB (1:1)	103	23	8.6

EML, emissive layer; CB, chlorobenzene; DCB, dichlorobenzene.

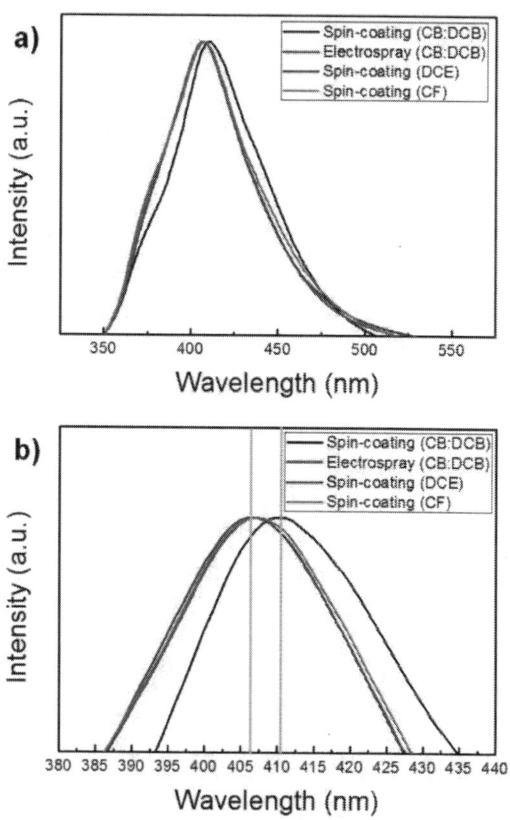

Figure 4 PL spectra of PVK films spin-coated or electrosprayed from the solvents (CB/DCB, DCE, or CF). (**a**) Normal view. (**b**) Enlarged view of (a). All spectra are normalized to their maximum value, and the excitation wavelength was fixed at 335 nm.

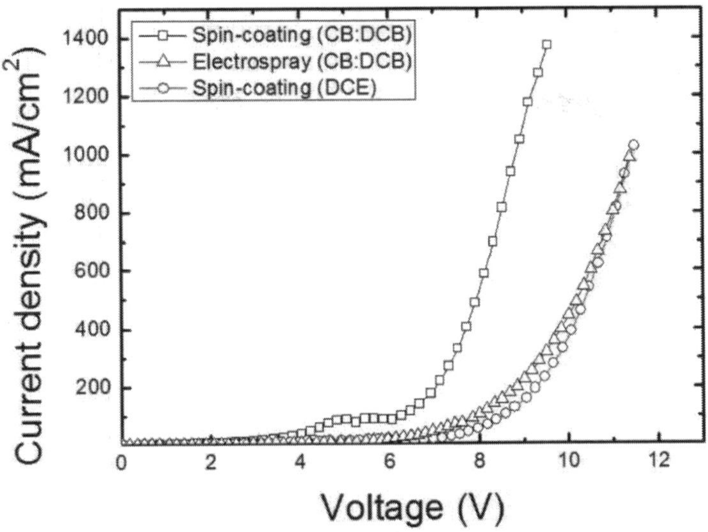

Figure 5 Current density-voltage curve of PVK devices using spin-coating and electrospray deposition (ITO/PEDOT:PSS/PVK/CsF/Al). The thin films using CB/DCB showed a better current density under the same conditions.

ACKNOWLEDGEMENTS

This research was supported by the Basic Science Research Program through the National Research Foundation of Korea (NRF) funded by the Ministry of Education, Science and Technology (2011-0012279) and (2011-0006268).

COMPETING INTERESTS

The authors declare that they have no competing interests.

AUTHORS' CONTRIBUTIONS

WH carried out the overall experiment. GX helped in the PL measurement. MC participated in the device fabrication. SMC and HC gave advice for this work. All authors read and approved the final manuscript.

REFERENCES

1. Tang CW, VanSlyke SA: Organic electroluminescent diodes. *Appl Phys Lett* 1987, 51: 913–915.

2. Burroughes JH, Bradley DDC, Brown AR, Marks RN, Mackay K, Friend RH, Burns PL, Holmes AB: Light-emitting diodes based on conjugated polymers. *Nature* 1990, 347: 539–541.

3. Baldo MA, O'Brien DF, You Y, Shoustikov A, Sibley S, Thompson ME, Forrest SR: Highly efficient phosphorescent emission from organic electroluminescent devices. *Nature* 1998, 395: 151–154.

4. Baldo M, Deutsch M, Burrows P, Gossenberger H, Gerstenberg M, Ban V, Forrest S: Organic vapor phase deposition. *Adv Mater* 1998, 10: 1505–1514.

5. Hebner TR, Wu CC, Marcy D, Lu MH, Sturm JC: Ink-jet printing of doped polymers for organic light emitting devices. *Appl Phys Lett* 1998, 72: 519–521.

6. Pardo DA, Jabbour GE, Peyghambarian N: Application of screen printing in the fabrication of organic light-emitting devices. *Adv Mater*2000, 12: 1249–1252.

7. Tseng S-R, Meng H-F, Lee K-C, Horng S-F: Multilayer polymer light-emitting diodes by blade coating method. *Appl Phys Lett* 2008, 93: 153308–153303.

8. Kopola P, Tuomikoski M, Suhonen R, Maaninen A: Gravure printed organic light emitting diodes for lighting applications. *Thin Solid Films*2009, 517: 5757–5762.

9. O'Regan M: Solution processed OLED displays: advances in performance, resolution, lifetime and appearance. *SID Symp Digest Tech Pap*2009, 40: 600–602.

10. Jin H, Sturm JC: Super-high-resolution transfer printing for full-color OLED display patterning. *J Soc Inf Display* 2010, 18: 141–145.

11. Saf R, Goriup M, Steindl T, Hamedinger TE, Sandholzer D, Hayn G: Thin organic films by atmospheric-pressure ion deposition. *Nat Mater*2004, 3: 323–329.

12. Ju J, Yamagata Y, Higuchi T: Thin-film fabrication method for organic light-emitting diodes using electrospray deposition. *Adv Mater* 2009, 21: 4343–4347.

13. Kim J-S, Chung W-S, Kim K, Kim DY, Paeng K-J, Jo SM, Jang S-Y: Performance optimization of polymer solar cells using electrostatically sprayed photoactive layers. *Adv Funct Mater* 2010, 20: 3538–3546.

14. Wilhelm O, Mädler L, Pratsinis SE: Electrospray evaporation and deposition. *J Aerosol Sci* 2003, 34: 815–836.

15. Rietveld IB, Kobayashi K, Yamada H, Matsushige K: Electrospray deposition, model, and experiment: toward general control of film morphology. *J Phys Chem B* 2006, 110: 23351–23364.

16. Jaworek A, Sobczyk AT: Electrospraying route to nanotechnology: an overview. *J Electrostat* 2008, 66: 197–219.

17. Lide DR: *Handbook of Organic Solvents*. Boca Raton: CRC Press; 1995.

18. Kim JW, Yamagata Y, Kim BJ, Higuchi T: Direct and dry micro-patterning of nano-particles by electrospray deposition through a

micro-stencil mask. *J Micromechanics Microengineering* 2009, 19: 025021.

19. Qian L, Bera D, Holloway PH: Electrophosphorescence from triplet excimers in poly-(N-vinylcarbazole). *Appl Phys Lett* 2007, 90: 103511–103513.

6

CHARACTERIZATION OF ELECTROCHROMIC PROPERTIES OF POLYANILINE THIN FILMS ELECTROPOLYMERIZED IN H_2SO_4 SOLUTION

Dang Hai Ninh[1], Tran Thi Thao[1], Pham Duy Long[2], Nguyen Nang Dinh[1]

[1]University of Engineering and Technology, Vietnam National University Hanoi, Hanoi, Vietnam

[2]Institute of Materials Science, Vietnamese Academy of Science and Technology, Hanoi, Vietnam

ABSTRACT

Polyaniline (PANI) onto indium-doped tin-oxide (ITO)-coated glass samples were prepared by electroopolymerization in 0.5 M H_2SO_4 solution. Structure and morphology characterization of the PANI films demonstrated that the films were grown onto ITO substrates in the form of polycrystalline microbelts separated by micropores. By analysing the UV-Vis absorption spectra of the PANI films, the energy bandgap was found to be approximately 2.75 eV. The PANI/ITO films exhibited a good reversible electrochromic display (ECD) performance when cycled in 0.1 M $LiClO_4$ + pro-py- lene carbonate. The response time of the ECD coloration was found to be as small as 15 s and the coloration efficiency was found to be 8.85 cm^2 ′ C^{-1}. After 100 cycles of the ECD performance, the cyclic voltammetry curve of the working electrode maintained unchanged. This demonstrates that the electropolymerized PANI films can be served as a good candidate for ECD applications, taking advantage of their excellent properties in terms of chemical stability.

KEYWORDS

Polyaniline, Thin Films, UV-Vis Absorption, Electrochromic Property

1. INTRODUCTION

The properties of the electrochromic display (ECD) of almost all the transition-metals oxides have been investigated [1] . These oxides films can be colored anodically (Ir, Ni) or cathodically (W, Ti). Together with inorganic materials, conducting polymers have also been intensively studied for polymer electrochromic devices (PECDs) [2]. Besides, conjugate polymers have been used for other applications such as polymer conductors [3] , electronic components [4] [5] , organic light-emitting diodes (OLED) [6] and solar cells (OSC) [7] . This is because conjugate polymers exhibit valuable advantages, such as thermal stability, low cost and easy preparation. Among the PECD materials, polyaniline (PANI) is widely used. There are some works focusing on PANI/ITO thin films for electrochemical supercapacitor applications. For example, the PANI films with different thicknesses and nanostructures were synthesized onto ITO substrates by using an electrodeposition technique using selenious [8] or sulfuric acid [9] . PANI has three main stable oxidation states: the fully reduced leucoemeraldine form, the 50% oxidized emeraldine form, and the fully oxidized pernigraniline form. Each of these states can exist in the form of its base or in the form of its protonated salt by subtracting or adding protons which does not modify its oxidation state. So that PANI is a multi-colour electrochromic material, such as transparent yellow for the leucoemeraldine form, green for the emeraldine salt form (ES), blue for the emeraldine base form (EB), and black for the pernigraniline form. With doping, the conductivity of PANI can reach a value from 10^{-10} to 10

$W^{-1} \cdot cm^{-1}$ [10] . The conductivity of PANI films mainly depends on the oxidation state, pH of the electrolyte and the type of dopant used for protonation [11] . PANI exhibits reverse electrochromic performance; it is able to switch from a transparent yellow state to a green one, according to reduction and oxidation processes [12] .

In this work, we report the investigation results on the structural, optical, morphological and electrochromic properties of PANI thin films prepared by electropolymerization in H_2SO_4 solution.

2. EXPERIMENTAL

Analytical reagent grade aniline and sulfuric acid (H_2SO_4) were used for the electropolymerization of PANI in an aqueous solution. 10 W indium-doped tin-oxide (ITO)-coated glass was used for substrates. Electrochemical processes were carried-out by using an AUTOLAB-Potentiostat PGS-30 electrochemical unit in a standard three-electrode cell, where PANI/ITO served as working electrode (WE), Ag/AgCl as reference electrode (AAE) and a platinum grid as counter electrode (CE). The electrolyte used for electropolymerization was composed of aqueous solution of 0.1 M aniline and 0.5 M dopant sulfuric acid. PANI thin films were electropolymerized by sweeping the potential between −0.20 V and +1.20 V/AAE for 10 numbers of scans. All the prepartion experiments were performed at room temperature without stirring. The as-deposited PANI films were dried in gaseous nitrogen and kept in a clean glove-box until use for further properties characterization. The electrochromic performance was carried out on the same PGS-30 potentiostat, using 0.1 M $LiClO_4$ + propylene carbonate ($LiClO_4$ + PC) solution for electrolyte.

The surface morphology was studied using a Hitachi "S-4800" field-emission scanning electron microscope (FE-SEM). X-ray diffraction analysis (XRD) was done on a Brucker "Advance-8D" X-ray diffractometer. Optical absorption measurements were carried using a JASCO ultraviolet-visible-near infrared (UV-Vis-NIR) "V-570" spectrophotometer. By using this photospectrometer, in situ transmittance spectra of PANI/ITO in $LiClO_4$ + PC vs. polarized potentials were recorded on the PANI/ITO films as the WE mounted into a modified electrochemical cell which was placed under the pathway of the laser beam. All measurements were executed at room temperature.

3. RESULTS AND DISCUSSION

The PANI films electropolymerization process in H_2SO_4 solution was strongly dependent on both the concentration of H_2SO_4, the scan rate and the range of the potentials used. Figure 1 shows typical cyclic voltammograms (CVs) of best samples obtained in the PANI films electropolymerization processes. The electrolyte contains 0.1 M aniline monomer with 0.5 M concentration of H_2SO_4. The CVs were recorded for 10 successive cycles at a scan rate of 50 mV/s for potential range from −0.20 to +1.20 V/AAE. A high redox peak appeared at ~1.15 V/AAE for the first cycle of CV. According to report in [13] , this peak is associated with the oxidation of aniline oligomers. We also observed the decrease in growth rate upon repeating the potential cycle, as in [14] the authors explained this phenomenon by the poor electrochemical activity of the PANI film formed on the electrode surface due to the difficulty in maintaining the protonated and doped structures of the polymers. The successive anodic scan shows the presence of three distinct peaks which are marked as "a", "b" and "c" attributing to the oxidation of fully reduced form of PANI (leucoemeridine) to its polaron form, oxidation of intermediate species, and final transition from delocalized polaronic state to bipolaron or quinoid form, respectively [14] . The potential of peak "a" is shifted toward higher anodic side with successive scanning could be due to the degradation of PANI deposited in the previous scan. After completing 10 cycles, greenish-colored thin films were deposited which confirms the formation of ES of PANI. The thickness (d) of the film measured on a profilometer (Deltak 6 M) was about 200 nm.

The crystalline structure of the PANI/ITO sample was checked by using an accessory for films with a small angle of the X-ray incident beam. For such a thin PANI film, some XRD patterns of the ITO substrate also appeared. Thus the XRD diagram shows both the diffraction peaks of ITO substrate and PANI film.

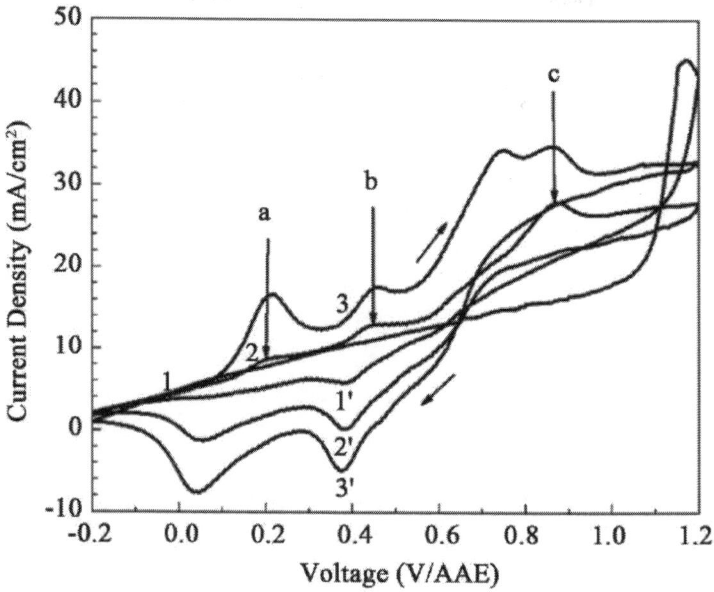

Figure 1. CV curves of PANI/ITO films made by electropolimerization in 0.1M aniline and 0.5 M H_2SO_4 solution by CV-cycling with a rate of 50 mV/s: (1-1')—First cycle; (2-2')—5-th cycle; and (3-3')—10-th cycle.

Indeed, in Figure 2 there are four diffraction peaks revealed which are quite consistent with the peaks of the atomic distances for a crystalline phase of In_2O_3—the host lattice of the ITO. Two intense bands observed approximately at 2q of 15.5° and 24.3° confirmed the synthesis of ES of PANI [15] .

The crystallinity of PANI thin film deposited in the presence of 0.5 M H_2SO_4 was found to be improved due to the change in surface morphology and compactness. The value of full width half maximum calculated for the peak at 2q = 24.3° was of as small as 0.06°. Indeed, we observed the change in the morphology as well as compactness for the PANI film deposited in 0.5 M concentration of H_2SO_4 by means of FE-SEM micrographs.

The morphology of the PANI films electropolymerized in the presence of 0.5 M concentration of dopant acid is shown in Figure 3. The microbelts of diameter approximately 200 nm are formed. The film was compact and fully deposited all over the substrate, this could be explained due to the high current flow from electrolyte to the electrode. Between the microbelts there are numerous pores served as the microcanals that constitute a factor favouring ions faster insert (extract) in (out of) the PANI films. This may improve both the response time and the performance efficiency of the electrochromic coloration.

The optical absorption spectra are often used to investigate the oxidation state of PANI. Figure 4shows the optical absorption spectra of a PANI film obtained during the electropolymerization at 0.5 M concentration of 0.5 M H_2SO_4 at a fixed scan rate of 50 mV/s. Three strong absorption peaks attributed at 328, 420 and 789 nm correspond to the formation of PANI in ES form. The peak observed at 317 nm is attributed to the transition of electrons from the highest occupied molecular orbital (HUMO) to the lowest unoccupied molecular orbital (LUMO) which is related to the p-p* electronic transition [16] . The other two absorption peaks observed at about 320 and 789 nm are associated with the polaron and bipolaron band transitions of PANI, respectively [16] . It is known that the polaron and bipolaron are the charge carriers in conducting PANI.

UV-Vis data can be used to estimate the energy bandgap, E_g, of the PANI films by using the following expression [17] .

$$\alpha h\nu = A\left(h\nu - E_g\right)^n \quad (1)$$

where h is Planck's constant, v is the frequency of the incident UV-Vis radiation, A is a constant and n is 1/2 for direct band semiconductors and 2 for indirect band gap semiconductors. As expected, best fits were obtained for n = 2 (indirect band).

A graph is plotted between $(ahn)^2$ or the square of (ahn) and ahn (as abscissa), a straight line is obtained. From the extrapolation of the straight line to $(ahn)^2 = 0$ axis one can determine the bandgap of the investigated sample. Thus from our experiments, the energy gap of

PANI films deposited by electropolymerization in the presence of 0.5 M concentration of H_2SO_4 was found to be of approximately 2.75 eV (Figure 5). This value is comparable to that for PANI prepared by redox polymerization of aniline using ammonium persulphate, $(NH_4)_2S_2O_x$, as dopant [18] .

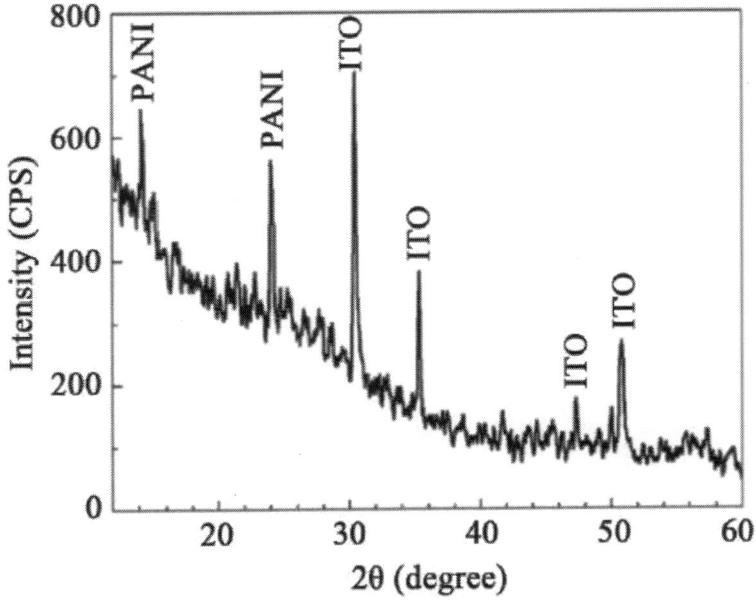

Figure 2. XRD pattern for a PANI thin film electropolymerized in the presence of 0.5 M concentration of H_2SO_4; d ~ 200 nm.

Figure 6 shows the cyclic voltammetry (CV) curve in 0.1 M $LiClO_4$ + PC of a PANI/ITO film, the CV spectra being recorded at the fifth cycle. Such a curve is typical of films prepared in our studies with a thickness of 200 nm. From this figure one can see the symmetrical shape of the CV spectra toward the coordinates (x = 0.5; y = 0.2). In the positive sweep direction (PSD) a peak of the anodic current density corresponding to a value of ca. 0.64 mA was obtained at a potential of 0.75 V/AAE. A negative value (-0.22 mA) of the peak in the negative sweep direction (NSD) was obtained at a potential of 0.35 V/AAE. The symmetrical CV proves a good reversibility of the processes of (ClO_4^-) ion insertion/extraction from the electrolyte into /out of the working electrode (PANI/ITO). During the oxidation, the ClO_4^- ions of $LiClO_4$/PC electrolyte solution were injected to the PANI thin film. During the reduction, the ClO_4^- ions of $LiClO_4$/PC electrolyte solution were excluded the PANI thin film. The overall reaction, involving ion insertion (extraction) in (out of) the PANI film to balance the charge, can be represented as follows [19] :

$$\underset{\text{(Light Green)}}{(ANI)n + ny(ClO_4^-)} \leftrightarrow \underset{\text{(Green)}}{[(ANI+)+y(ClO_4^-)]n} + nye^- \quad (2)$$

where n is the number of repeated units and y is the stoichiometric number of the counter ion.

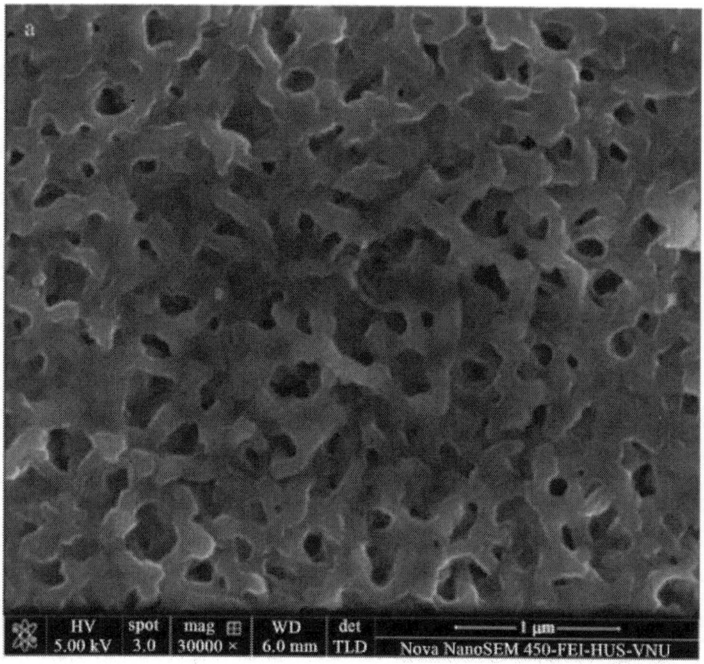

Figure 3. SEM of as-deposited PANI thin film (d ~ 200 nm) onto ITO substrate in H_2SO_4 solution.

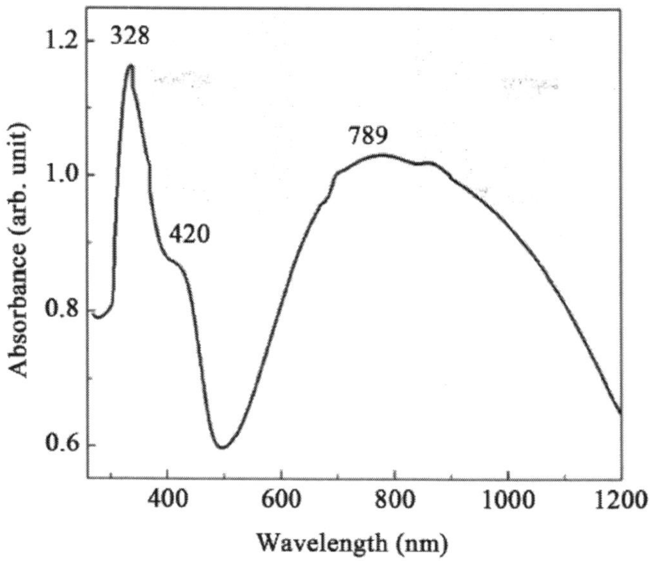

Figure 4. Absorption spectra of a PANI/ITO thin film electropolymerized in the presence of 0.5 M concentration of H_2SO_4.

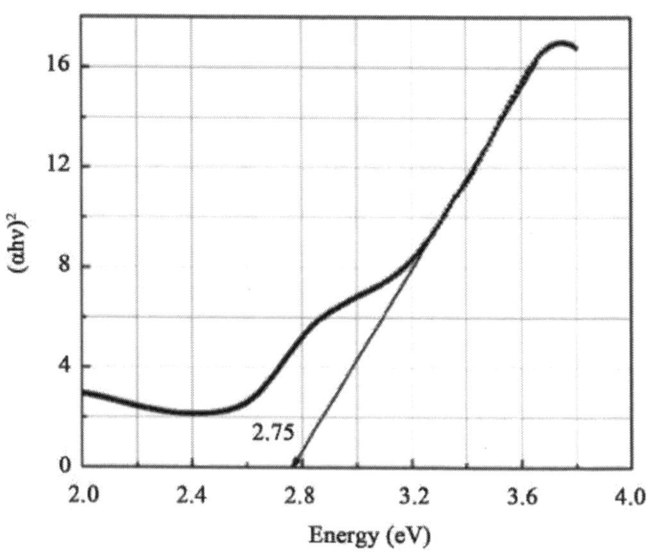

Figure 5. Bandgap determination of PANI film electropolymerized in 0.5 M H_2SO_4 solution.

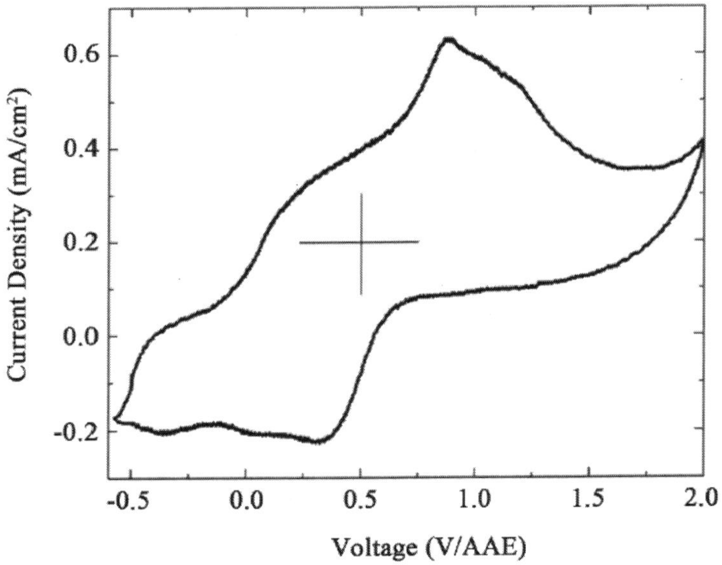

Figure 6. Cyclic voltammetry of PANI/ITO/Glass in 0.1 M $LiClO_4$ + PC; the scanning rate is of 50 mV/s.

To study the electrochromic performance, a device with the structure of Glass/ITO/Electrolyte/PANI/ITO/ Glass (abbreviated to Pani-ECD) was prepared.

The $LiClO_4$ + PC electrolyte was dropped onto the working electrode with a 1 × 1 cm^2 active area and thickness of the electrolyte was fixed at the same value of the thickness of the Scotch tape (i.e. ~50 μm).The PANI/ ITO and ITO/Glass electrodes of the device were connected to WE and CE of the PGS-30 with a polarized potential of −0.5 V and + 1.5 V. For the device with a 200 nm-thick PANI film, the in situ transmission spectra obtained during electrochromic performance are given inFigure 7. The completely coloured (curve "a") and bleached (curve "b") states occurred considerably fast, after approximately 15 sec and 5 sec, respectively. The PANI/ITO has the maximum optical transmittance variation, ΔT (%), which was larger 40% at 530 nm.

To evaluate the electrochromic coloration efficiency (h) we used a well-known expression relating the efficiency with the optical density, consequently the transmittances of coloration (T_c) and bleaching states (T_b), and the insertion charge (Q) are as follows [20] :

$$\eta = \frac{\Delta OD}{Q} = \frac{1}{Q}\ln\left(\frac{T_b}{T_c}\right) \quad (3)$$

In our experiments Q = 0.14 mC ′ cm^{-2}. At a wavelength of 530 nm, T_b = 62% and T_c = 18%, thus the coloration efficiency was found to be of ~8.85 cm^2 ′ C^{-1}. In the visible range of wavelengths all the values of h found are comparable to those for PANI/ITO/Glass that was electropolymerized in 2 M HCl solution [21] and a little higher than those for PANI/ITO/Glass prepared by electrochemistry [22] .

To investigate the durability of the PANI films, a PANI/ITO/Glass of 1.0 cm^2 in size was measured in 0.1M $LiClO_4$ + PC. The scanning potentials range was from −0.5 to + 2.0V/AAE and a number of the cycles was chosen as large as 100 cycles. The CV curves recorded from the first to hundredth cycle are shown in Figure 8. From the fifth to tenth cycle, in both the PSD and NSD the current density in absolute value was found to slightly increase; it then slowly decreased. After 100 cycles, the CV curve was maintained unchanged and the current density of the CV peak lowered to a value of ~90% of the initial value (at the saturation coloration state, i.e. at the tenth cycle of the cyclic voltammetry). This demonstrates that the ClO_4^- insertion (extraction) into (out of) the porous PANI films could be easily performed. For the TiO_2/ITO films deposited by the sol-gel technique, the time to get a saturated state of coloration was as large as 45 min for a sample size of 1 cm^2 [23] . In the present work, the PANI/ITO was coloured very rapidly for a sample of the same size. The saturated coloration was reached about 15 sec after a negative potential of -1.20 V/AAE was applied to the WE in the 0.1 M $LiClO_4$ + PC electrolyte. A deep green colour was observed in the coloration state and a light green (bleached state) was obtained after less than 5 sec.

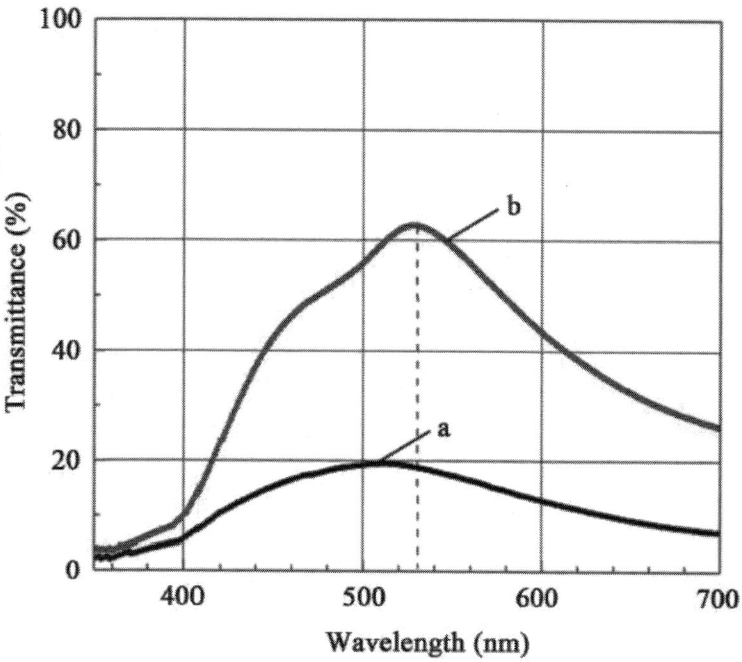

Figure 7. Transmittances of the Pani-ECD device. The oxidizing potential and reducing potential were set +1.0 V and −0.5 V, respectively. Curve "a" is the colored state and curve "b" is bleached state.

4. CONCLUSIONS

PANI/ITO/Glass thin films were electropolymerized in the presence of 0.5 M concentration of H_2SO_4. The PANI films were formed in polycrystalline microbelts separated each other by micropores. From UV-Vis data of the films, the energy bandgap of the PANI films was estimated, as approximately 2.75 eV. Electrochromic performance of the Pani-ECD device was carried out in 0.1 M $LiClO_4$ + propylene carbonate electrolyte, and a good reversible coloration and bleaching process was obtained. The response time and the coloration efficiency of the coloration of PANI/ITO films were found to be 15 s and 8.85 cm^2 ′ C^{-1}, respectively.

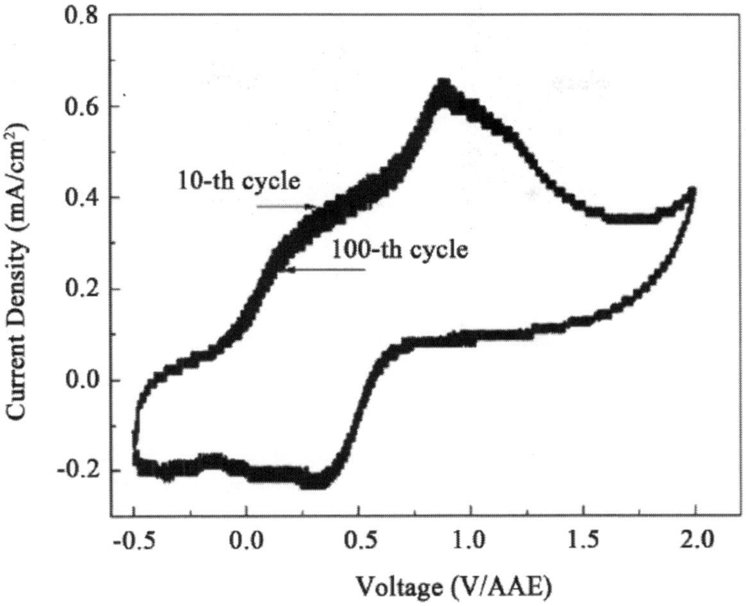

Figure 8. Testing of the durability of an ECD device with a structure of Glass/ITO/PANI/$LiClO_4$ + PC/ITO/Glass; the scanning rate is of 50 mV/s.

The results of the ECD performance for 100 cycles showed that electropolymerized PANI thin films possess a long durability. Thus polyaniline thin films can be served as a good candidate for ECD applications, taking advantage of their excellent properties in terms of chemical stability.

ACKNOWLEDGEMENTS

This research is funded by Vietnam National Foundation for Science and Technology (NAFOSTED) under grant number 103.02-2013.39.

REFERENCES

1. Granqvist, C.G. (1995) Handbook of Inorganic Electrochromic Materials. Elsevier, Amsterdam.

2. Yildirim, A., Tarkuc, S. and Toppare, M.A.L. (2008) Syntheses of Electroactive Layers Based on Functionalized Anthracene for Electrochromic Applications. Electrochimica Acta, 53, 4875-4882.

3. Crispin, X., Jakobsson, F.L.E., Crispin, A., Grim, P.C.M., Andersson, P., Volodin, A., van Haesendonck, C., Van der Auweraer, M., Salaneck, W.R. and Berggren, M. (2006) The Origin of the High Conductivity of Poly(3,4-ethylenedioxy-thiophene)-poly(styrenesulfonate) (PEDOT-PSS) Plastic Electrodes. Chemistry of Materials, 18, 4354-4360.

4. Sirringhaus, H., Kawase, T., Friend, R.H., Shimoda, T., Inbasekaran, M., Wu, W. and Woo, E.P. (2000) High-Resolution Inkjet Printing of All-Polymer Transistor Circuits. Science, 290, 2123-2126.

5. Katz, H.E. and Huang, J. (2009) Thin-Film Organic Electronic Devices. Annual Review of Materials Research, 39, 71-92.

6. Parker, I.D. (1994) Carrier Tunneling and Device Characteristics in Polymer Light-Emitting Diodes. Journal of Applied Physics, 75, 1656-1666.

7. McCullough, R.D. and Ewbank, P.C. (1998) Regioregular, Head-to-Tail Coupled Poly(3-alkylthiophene) and Its Derivatives. In: Skotheim, T.A., Ed., Handbook of Conducting Polymers, Marcel Dekker, New York, Chapter 9, 225-258.

8. Amarnath, C.A., Chang, J., Lee, J., Kim, D.Y., Mane, R.S., Han, S.-H. and Sohn, D. (2008) Direct Polymerized Polyaniline Nanostructures on Modified Indium-Tin Oxide Surface for Electrochemical Supercapacitors. Electrochemical and Solid-State Letters, 11/10, A167-A169.

9. Inamdar, A.I., Kim, Y.S., Sohn, J.S., Im, H., Kim, H., Kim, D.-Y., Kalubarme, R.S. and Park, C.J. (2011) Supercapacitive

Characteristics of Electrodeposited Polyaniline Thin Films Grown on Indium-Doped Tin-Oxide Substrates. JKPS, 59, 145-149.

10. Salavagione, H.J., Sanchis, C. and Moralln, E. (2007) Friendly Conditions Synthesis of Platinum Nanoparticles Supported on a Con-Ducting Polymer: Methanol Electrooxidation. Journal of Physical Chemistry C, 111, 12454-12460.

11. Focke, W.W., Wnek, G.E. and Wei, Y. (1987) Influence of Oxidation State, pH, and Counterion on the Conductivity of Polyaniline. The Journal of Physical Chemistry, 91, 5813-5818.

12. Ashwini, B., Priyanka, U.L., Mahapatra, S.K., Kulkarni, S.K. and Chaure, N.B. (2014) Electropolymerization of Polyaniline Thin Films. High Performance Polymers, 26, 641-646.

13. Hong, S.Y. and Park, S.M. (2005) Electrochemistry of Conductive Polymers 36. pH Dependence of Polyaniline Conductivities Studied by Current-Sensing Atomic Force Microscopy. The Journal of Physical Chemistry B, 109, 9305-9310.

14. Stilwell, D.E. and Park, S.M.J. (1988) Electrochemistry of Conductive Polymers II. Electrochemical Studies on Growth Properties of Polyaniline. Journal of the Electrochemical Society, 135, 2254-2262.

15. Chaudhari, H.K. and Kelkar, D.S. (1996) X-Ray Diffraction Study of Doped Polyaniline. Journal of Applied Polymer Science, 62, 15-18.

16. Abdulla, H.S. and Abbo, A.I. (2012) Optical and Electrical Properties of Thin Films of Polyaniline and Polypyrrole. International Journal of Electrochemical Science, 7, 10666-10678.

17. Tauc, J. (1974) Amorphous and Liquid Semiconductors. Plenum, New York, 159.

18. Joshi, G.P., Saxena, N.S., Sharma, T.P., Dixit, V. and Misra, S.C.K. (2003) Bandgap Determination of Chemically Doped Polyaniline Materials from Reflectance Measurements. Indian Journal of Pure & Applied Physics, 41, 462-465.

19. Shen, K.Y., Hu, C.W., Chang, L.C. and Ho, K.C. (2012) A Complementary Electrochromic Device Based on Carbon Nanotubes/Conducting Polymers. Solar Energy Materials & Solar Cells, 98, 294-299.

20. Monk, P.M.S., Mortimer, R.J. and Rosseinsky, D.R. (1995) Electrochromism: Fundamentals and Applications. VCH, Weinheim-New York-Basel-Cambridge-Tokyo.

21. Liu, C.Y., Chou, J.C., Liao, Y.H., Yang, C.J., Huang, C.J., Cheng, T.Y., Hu, J.E. and Chou, H.T. (2013) The Investigation of the Electrochromic Characteristics for the PANI Thin Film by Cyclic Voltammetry and Potentiostatic Method. Proceedings of the World Congress on Engineering, Vol. 2, London, 3-5 July 2013, 999-1003.

22. Kang, J.H., Oh, Y.J., Peak, S.M., Hwang, S.J. and Choy, J.H. (2009) Electrochromic Device of PEDOT-PANI Hybrid System for Fast Response and High Optical Contrast. Solar Energy Materials & Solar Cells, 93, 2040-2044.

23. Dinh, N.N., Oanh, N.T.T., Long, P.D., Bernard, M.C. and Hugot-Le Goff, A. (2003) Elecrochromic Properties of TiO_2 Anatase Thin Films Prepared by Dipping Sol-Gel Method. Thin Solid Films, 423, 70-76.

7

PREPARATION AND CHARACTERIZATION OF SUBSURFACE SILVER PARTICULATE FILMS ON POLYMER BLENDS OF POLYSTYRENE/POLY (2-VINYLPYRIDINE)/POLY(VINYLPYROLLIDONE)/ POLY(4-VINYLPYRIDINE)

Pratima Parashar Pandey*

Department of Materials Science, Mangalore University, Mangalagangotri, India

ABSTRACT

Silver particulate thin films on softened polymer blends of Polystyrene (PS)/Poly(2-vinyl pyridine) (P2VP), PS/Poly(4-vinylpyridine) (P4VP), and Poly(vinylpyrollidone) (PVP)/P4VP at a rate of 0.4 nm/s held at a temperature of 457 K in vacuum of 8×10^{-6}Torr by evaporation are deposited. These silver films were characterized by their electrical behavior, optical properties and Scanning electron microscopy (SEM). Silver films deposited on softened PS, and PVP give rise to a very high room temperature resistance approaching that of the substrate resistance due to the formation of a highly agglomerated structure. On the other hand, silver films on softened P2VP and P4VP gives rise to a room temperature resistance in the range of tens to a few hundred MΩ/□ which is desirable for device applications. Silver films on the composites of PS/P2VP, PS/P4VP and PVP/P4VP show resistances at room temperature. The optical and plasmonic response of Ag nanoparticles onto thin layers of blends shows encapsulation of nanoparticles. The electrical properties and SEM of silver nanoparticles on the thin layers of polymer blends indicate the formation of much smaller, narrower dispersion and wide size distribution.

KEYWORDS

Nanoparticles, Polymer Blends, Thin Films, Vapor Deposition, Electrical Properties, Optical Properties, SEM

1. INTRODUCTION

Polymer/inorganic nanocomposites have been of great interest in recent years, not only for the novel properties of the nanocomposite materials but also for the continuously growing demand for the miniaturization of electronics components, optical detectors, chemical and biochemical sensors and devices [1] .

Polymer matrices have been frequently used as particle stabilizers in synthesis of metal colloids since these prevent agglomeration of the particles. Within the past decade, incorporating silver nanoparticle into a polymer matrix is more interesting because the resulting nanocomposites exhibit applications in catalysis, drug delivery, wound dressing, antimicrobial activity and optical information storage [2] -[8] .

It is difficult to disperse silver nanoparticle homogeneously into a polymer matrix by ex situ methods because of easy agglomeration of nanoparticles [9] . At present, it is possible to obtain nanoparticles of different shape and size in nano structured polymeric environment where various polymeric systems and different approaches can be used.

Numerous methods used toxic and potentially hazardous reactants. Increasing environmental concerns is resulted in an attempt to adopt eco friendly methods.

One of the simplest and eco friendly techniques to form such particulate structures, which are generally known as island or discontinuous metal films, is through vacuum evaporation of metal on to a dielectric substrate by stopping the deposition at a very early stage. The temporal instability exhibited by island films even in vacuum is attributed to mobility of islands followed by coalescence. Further, these films get oxidized when they are exposed to atmosphere. The oxidation of islands causes an irreversible increase in electrical resistance [10] . An interesting sub-surface particulate structure formation was reported when certain inorganic materials are deposited on to softened polymer substrates [11] - [14] . The morphology and formation of such structures depend on thermodynamic as well as deposition parameters [13] [14] . The use of softened polymer substrates provides the unique possibility of easily controlling the viscosity of the substrate to form a subsurface discontinuous silver particulate films. The morphology of sub-surface particulate structures also depends upon polymer metal interaction [14] - [17] . The reported method is evaporation of silver on polymer substrate at high temperature and in vacuum of the order of 10^{-6}Torr. The ability to precisely tailor and optimize the nanocomposite structure creates opportunities for a wide range of applications.

Blending of polymers provide the requisite matrix so that the property of PS and PVP can be combined with P2VP and P4VP. By varying the blend compositions suitable polymer composite exhibited the desired results. The subsurface silver films could be deposited on PS/P2VP (50:50), PS/P4VP (50:50, 25:75) and PVP/P4VP (50:50) which yield room temperature resistances in the range of a few tens to a few hundred M/ desirable for device applications.

2. EXPERIMENTAL

Poly(4-vinyl pyridine), Poly(2-vinylpyridine) and Poly (vinylpyrollidone) used in this study, were procured from Sigma-Aldrich Chemicals Pvt. Ltd and Polystyrene from Alfa-Aesar (A Johnson Mathley Company) respectively. The molecular weights of P4VP, P2VP and PS are 60,000, ~37,500 and 100,000, respectively. The structure of P4VP, P2VP, PS and PVP are (a), (b), (c) and (d) respectively, as follows:

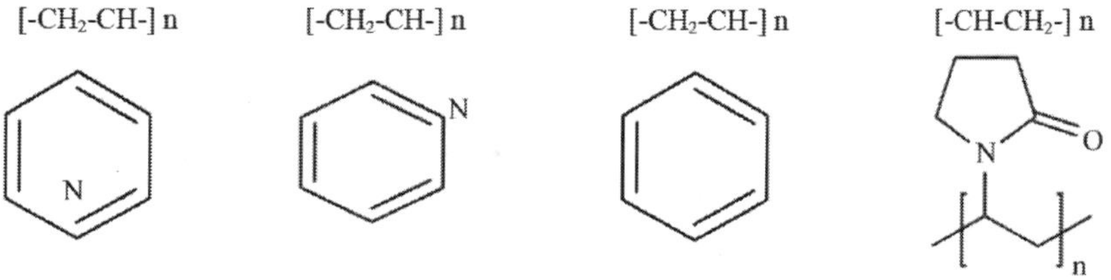

Polymer blends (PS/P2VP/P4VP) were prepared through solution blending by mixing in a common solvent, Dimethylformamide (DMF). Blends of PS/P4VP with different compositions {PS (w)/P4VP (w) = 0:100; 25:75, 50:50; 75:25; 100:0} were prepared. 1 g of the total polymers at different ratios was dissolved in 20 ml of DMF at room temperature. Composite of PS/P2VP with different compositions {PS (w)/P2VP (w) = 0:100; 50:50; 100:0} were prepared. An amount of 0.5 g of the total polymers at different ratios, were dissolved in 5 ml of DMF at room temperature.

Polymer blends of PVP/P4VP were prepared through solution blending by mixing in a common solvent, water. Blends of PVP/P4VP with different compositions {PVP (w)/P4VP (w) = 0:100; 50:50; 100:0} were prepared. 1 g of the total polymers at different ratios was dissolved in 20 ml double distilled water at room temperature.

Thin films of homopolymers and their composite of approximately 5 μm thickness were solution cast on a glass slide pre-coated with silver contacts with a gap of 1 cm × 1 cm for electrical studies. Silver films of various thick- nesses were deposited on these substrates held at 457 K in a vacuum better than 8×10^{-6} Torr. The glass transition temperature of P4VP, P2VP, PVP and PS are 410, 357, 379 and 373 K, respectively. Therefore, polymer substrates are softened at 457 K and sufficient fluidity is ensured. A chromel-alumel thermocouple was used to measure the substrate temperature by clamping it to the substrate surface holding the film. Source to substrate distance was maintained at 20 cm. A Telemark quartz crystal monitor (Model 850) was used to measure the deposition rate, as well as the overall film thickness. The deposition rate used was 0.4 nm/s for all the films. Resistance measurements were carried out in-situ, using a Keithley electrometer model 617. The films were annealed at the deposition temperature for 1 hour before cooling them to room temperature.

Optical absorption spectra of the silver particulate films were obtained on a Shimadzu Uv-Vis-NIR spectrophotometer model SHIMADZU UV 3101 PC.

Scanning electron microscopy (SEM) measurements were carried out on a Leica Cambridge Instruments Stereoscan 440 SEM with image processing software.

3. RESULT AND DISCUSSION

We need the compatibility of blends at higher temperature to ensure formation of polymer composite above their glass transition temperature. Blending of polymers provide the requisite matrix so that the property of PS and PVP can be combined with P2VP and P4VP. By varying the blend compositions suitable polymer composite exhibited the desired results. The compatibility of the solution cast PS/P4VP and PS/P2VP blends was studied by using Dilute Solution Viscometry (DSV), Differential Scanning Calorimetry (DSC), Fourier transformation infrared spectroscopy (FTIR) and Scanning Electron Microscopy (SEM). The blends were found to be compatible at higher temperature [15] [18] . The subsurface silver films could be deposited on PS/P2VP (50:50), PS/P4VP (50:50, 25:75) and PVP/P4VP (50:50) which yield room temperature resistances in the range of a few tens to a few hundred M/ desirable for device applications.

In-situ electrical resistance measurements yield about structure of discontinuous silver particulate films deposited on polymer composites. Table 1 presents the resistance data for the silver films of various thickness deposited on different polymer blend substrates held at 457 K with a rate of 0.4 nm/s. It is clear that room temperature resistance of silver films deposited at 457 K is very high for PS and PVP approaching that of the substrate resistances value where as a few mega ohms resistances are measured for P2VP and P4VP composites. This shows silver films are particulate in nature hence amounts for high resistances even for high thickness up to 200 nm. It may be due to silver deposited is dispersed in softened polymer composites and formed a subsurface particulate within the polymer matrix [14] . Negative temperature coefficient of resistances (TCR) was observed for all the films which is the characteristic of particulate film. Negative TCR is the indication of highly agglomerated sub- surface structure as observed for PS and PVP for any thickness and PS/P4VP (75:25) for 50 nm. But the magnitude of TCR is small for higher thickness films on PS/P4VP (75:25) for 150 nm, PS/P2VP (50:50) for 100, 150 and 200 nm and PVP/P4VP for 86 nm thick films. TCR for all such films was found to be zero at lower temperatures.

The dc resistances of the particulate film are exponentially dependent on inter-cluster separation or tunnelling length. Therefore, observed high resistances were attributed to very large inter-spacing and bigger cluster sizes for PS which is reduced with the decreasing content of PS in the blend and increasing thickness of film. Overall resistance of film deposited on composites decreases with increase in thickness of silver films. Similar behaviour is reported that the electrical conductivity of composites is increased with high silver loading (30% - 80%) [19] . Pyridine ring with Nitrogen atom in the structure of P2VP and P4VP causes the change in the morphology of these films as compared to PS and PVP.

Figures 1-3 show the SEM pictures of silver films deposited on PS/P2VP/P4VP composites. It is clear from Figures 1[1(a), 2(a), 3(a)] that size and inter-spacing in case of individual polymer for thickness of

100 nm is 109, 68 and 139 nm for P2VP, P4VP and PS, respectively. These are in agreement with their electrical properties. Blending of polymers resulted in a requisite polymer matrix which shows modification in the morphology of films and hence improvement in tunnelling length in the film deposited on polymer composites. Blending of P2VP and P4VP with PS is expected to provide a polymer matrix where the size of silver clusters and inter- cluster separation can be modified because of dispersion, size distribution and impregnation depth results from the natures of polymeric hosts [20] [21] .

Figure 2 clearly shows the particle sizes and inter-separation for PS/P4VP composites. The particle size for 50 nm films [2(a), 2(b)] on P4VP and PS are 58 and 92 nm, respectively. SEM 3a, 4a show the particle size distribution for 50 and 95 nm thick silver films deposited on PS/P4VP, 50:50. The data fit into a log normal distribution for both the cases. The average size, a and geometric standard deviation, σ_a are 79.4 nm and 1.2, respectively for the 50 nm film whereas the corresponding values for the 95 nm film are 95.4 and 1.4. A closer look at the morphology of silver nanoparticles deposited on PS/P4VP, 50:50 in SEM. 3a & 4a, clearly show that particle size increases with the amount of silver deposited. The shape of the nanoparticles changes from near spherical particles to irregular ellipsoidal particles. The average size of the particle increases from 79.4 to 95.4 nm and the size distribution expands from 50 - 110 nm to 60 - 160 nm which results in improvement of tunnelling effect in PS/P4VP, 50:50 [21] . It is evident that increase in size distribution decreases the inter-separation of silver clusters in this blend. This fact may be regarded as a consequence of the size as well as inter-separation evolution of nanoparticles during the ongoing deposition process. SEM pictures 5a & 6a show the particle size distribution for 50 nm and 150 nm thickness films deposited on PS/P4VP, 75:25. The average size, a and geometric standard deviation, σ_a are 88.6 nm and 8.5 for the 50 nm film and 96.7 and 4 for 150 nm film, respectively. It is evident from the figure that distribution of size increased from 50 to 120 nm to 60 to 140 nm. Such dispersion of silver nanoparticle within the PS/P4VP, 75:25 leads to better electrical behaviour [21] . Such electric behaviour is not observed even for 300 nm silver particulate film on PS [22] . Hence, silver particulate films on PS/P4VP, 75/25 at low volume fraction of silver consist of isolated and widely dispersed nanoparticles. But systematic and controlled increase of silver volume in the PS/P4VP, 75:25 matrixes produced interesting result [21] .

Table 1. Resistance data for silver films deposited on polymers blends held at 457 K at the rate of 0.4 nm/s.

Silver film thickness (in nm)	Resistances (MΩ/□)																	
	On PS		On PS/P2VP 50:50		On P2VP		On PVP		On PVP/P4VP 50:50		On PS/P4VP 25:75		On PS/P4VP 50:50		On PS/P4VP 75:25		On P4VP	
	457 K	300 K	457 K	300 K	457 K	300 K	457 K	300 K	457 K	300 K	457 K	300 K	457 K	300 K	457 K	300 K	457 K	300 K
50	302	>1000	-		-		15	>1000			6.2	47.1	15.9	159.4	214	>1000	1.9	27.1
85	-		-		-		-		7.2	113	3.2	14.2			-		-	
95	-		-		-		-		-		-		2.9	30.1	-		-	
100	-		47	115	12.28	32.6	-		-		-		-		-		-	
150	54	>1000	11.5	42	9.73	26.1	-		-		-		-		14.8		248.9	
200	-		10.35	30.1	-		-		-		-				-		-	

Figure 1. SEM of silver films deposited P2VP, P4VP and PS for 100 nm.

Figure 3 indicates modification in silver particulate films deposited on PS/P2VP composites. The particle sizes measured from respective SEM pictures [1(a), 2(a), 3(a)] fit into a log normal distribution for all the cases. The average size, a and geometric standard deviation, σ_a are 205 nm and 4; 109 nm and 9, 129 nm and 3, respectively, for silver films on PS, P2VP and their blend (50:50). The size distribution indicates that the particle size varies from 100 to 400 nm, 60 to 180 nm and 60 to 200 nm for silver films on PS, P2VP and their blend (50:50). It is clear from the figure that the size and inter-separation of silver clusters is wide in PS whereas size as well as inter- separation is less in P2VP [14] . As a result, PS do not show the desired electrical conductivity [22] . Blending of P2VP with PS modifies size, size distribution and inter-separation of silver particles on their blend PS/P2VP, 50:50 which results in improvement of tunnelling effect in the blend and the blend shows desired electrical behaviour [15] . SEM pictures [1(b), 2(b), 3(b)] show the particle size distribution for 100, 150 and 200 nm thickness films deposited on PS/P2VP, 50:50. The average size, a and geometric standard deviation, σ_a are 94.8 nm, 1.5 and 100 nm, 5 and 129 nm, 3 for the 100, 150 and 200 nm film, respectively. It is evident from the figure that size distribution varied from 55 - 125, 65 - 135 to 70 - 200 nm for the 100, 150 and 200 nm films, respectively. Such dispersion of silver nanoparticle within the PS/P2VP, 50:50 leads to better electrical behaviour [15] . This electric behaviour is not observed even for 300 nm silver particulate films on PS [22] . Hence, silver particulate films on PS/P2VP, 50:50 at low volume fraction of silver consist of isolated and widely dispersed nanoparticles. But systematic and controlled increase of silver volume in the blend matrixes has shown increase in the size of silver clusters [15] .

Silver nanocomposite film exhibits characteristic UV absorbance spectrum at a wavelength of 430 nm, due to the surface plasmon resonance of nanosized silver [23] - [26] . These nanoparticles exhibit unique optical properties originating from the characteristic surface plasmon by the collective motion of conduction electrons [25] [26] .

Thus, the formation of silver nanoparticles can also be confirmed by UV/VIS absorption spectrum of composite films [27] . Spectral position, half width and intensity of the plasmon resonance strongly depend on the particle size, shape and the dielectric properties of the particle material and the surrounding medium [27] . Thus, the type of metal and the surrounding dielectric medium play a significant role in the excitation of particle plasmon resonance (PPR). The sensitivity of PPR frequency to small variations of these parameters can be exploited in various applications [28] . The differences in dispersion, size distribution and impregnation depth results from the differing natures of the polymeric hosts [16] .

Figure 4 shows the optical absorption spectra recorded for 100 nm silver films deposited on polymeric blends held at 457 K at the deposition rate of 0.4 nm/s. Silver particles embedded in PS/P2VP blends observed a shift of the resonance position to higher wavelength (red shift) which were correlated with changes of particle sizes and inter-separation in silver clusters. It is clearly seen (Figure 4) that the plasmon resonance peak shifts towards the longer wavelength side for the PS/P2VP, 50:50 (435) as compared to pure polystyrene (429 nm). Also, there is increase in intensity of absorbing peaks which signify the decrease of particles size with the incorporation of P2VP into PS [24] . P2VP exhibits two peaks (441, 606 nm). It is interesting to note that PS/P2VP, 50:50 also shows an additional absorption band at higher wavelength (616 nm). The possible explanation is that silver nanoparticles are in a highly aggregated state leading to coupling of the plasmon vibrations between neighbouring particles [29] . Similar results were found for silver particulate films of 150 and 200 nm films on PS/P2VP 50:50 blends. The increase in particle size of silver nanoparticles are due to increasing thickness of silver particulate films on the blends (Figure 3) [30] .

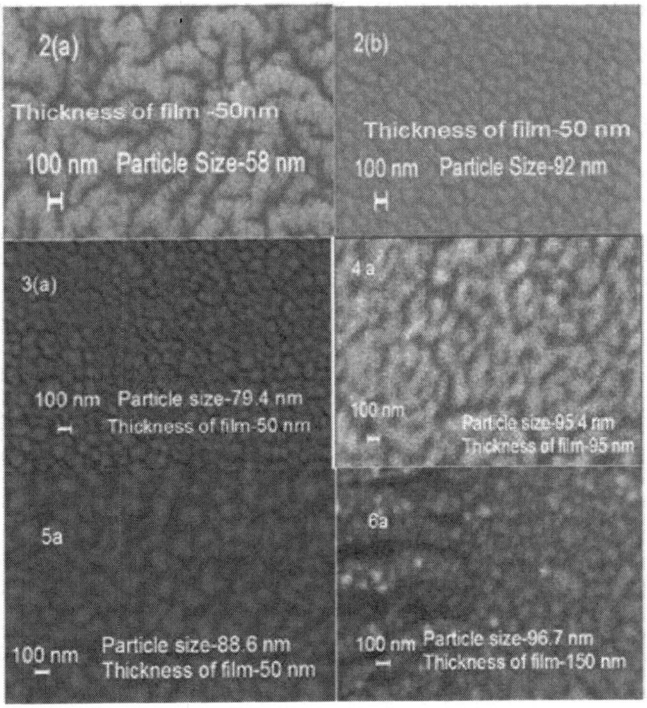

Figure 2. SEM of silver films on PS/P4VP composites for various thicknesses.

Figure 3. SEM of silver films on PS/P2VP composites for various thicknesses.

Figure 5 shows the optical absorption spectra recorded for 50 nm silver films deposited on polymeric blends held at 457 K at the deposition rate of 0.4 nm/s. For silver particles embedded in PS/P4VP blends, a shift of the resonance position to higher wavelength (red shift) was found, which were correlated with changes of particle sizes and inter-separation in silver clusters. It is clearly seen (Figure 5) that the plasmon resonance peak shifts towards the longer wavelength side as compared to pure polystyrene (485.5 nm) with increasing percentage of P4VP in the blends. Also, there is increase in intensity of absorbing peaks which signify the decrease of particles size with the incorporation of P4VP into PS [25] .

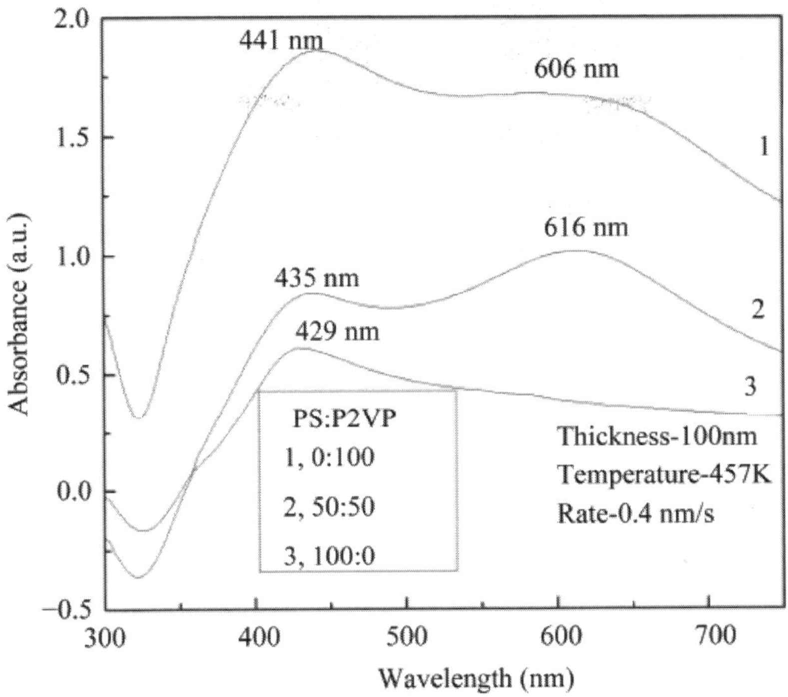

Figure 4. Optical absorption spectra for 100 nm silver particulate film deposited on PS/P2VP blends.

Figure 6 shows the optical absorption spectra recorded for 52 & 86 nm silver films deposited on polymeric blends of PVP/P4VP held at 457 K at the deposition rate of 0.4 nm/s. Silver particles embedded in PVP/P4VP blends observed a shift of the resonance position to higher wavelength (red shift) which were correlated with changes of particle sizes and inter-separation in silver clusters. It is clearly seen (Figure 6) that the plasmon resonance peak shifts towards the longer wavelength side for the PVP/P4VP, 50:50 (653) as compared to pure PVP (611 nm) for silver film thickness of 86 nm. Thus, the formation of silver nanoparticles is confirmed by UV/VIS absorption spectrum of composite films [25] .

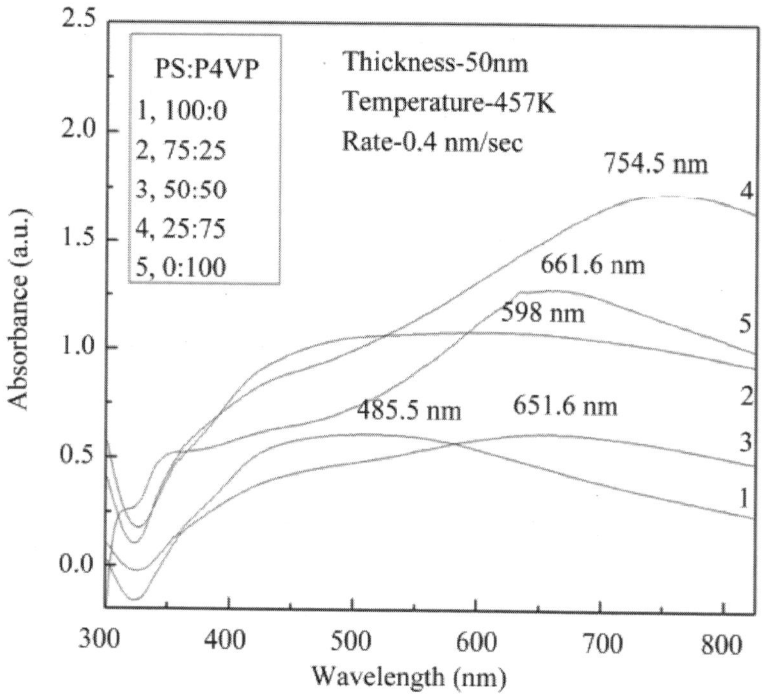

Figure 5. Optical absorption spectra for 50 nm silver particulate films deposited on PS/P4VP blends.

4. CONCLUSION

Deposition of silver on polymer blends coated substrate held at 457 K provides an approach to produce stable island films with reasonable control over their electrical resistance. Higher thickness films show almost zero TCR near room temperature, a desirable property for most of the devices. Low thickness films show a negative TCR, characteristic of island film. Silver particulate films deposited on PS/PVP/P2VP/P4VP blends show better electrical properties compared to pure PS and PVP. The size distribution and dispersion of silver nanoparticles is found to be dependent on the nature of the polymer host and thickness of particulate films. With the addition of P4VP, P2VP and amount of silver, morphology of the silver particulate films on PS/P2VP (50:50), PS/P4VP

(50:50, 75:25) and PVP/P4VP (50:50) could be modified to give the desired electrical results. It is an approach towards obtaining low cost, stable thin silver films on polymer composites for device applications.

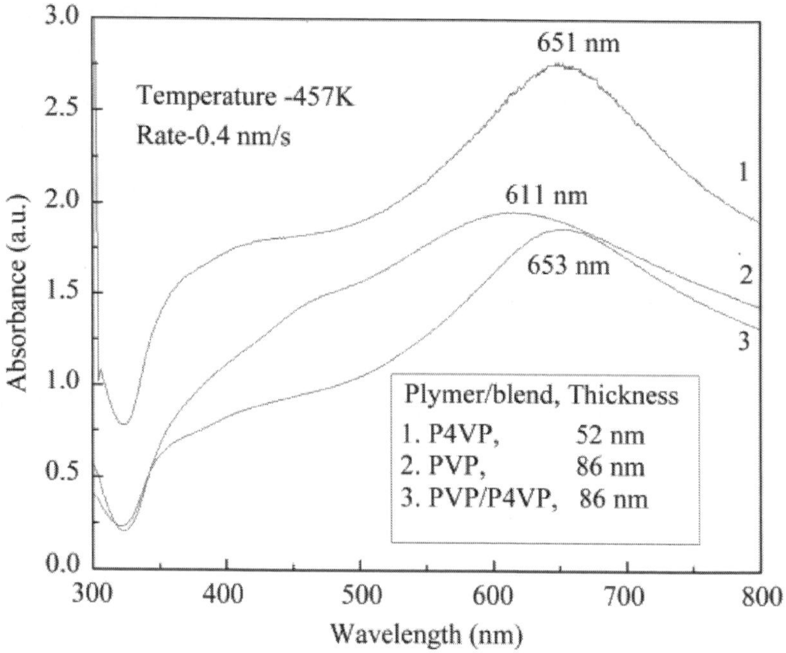

Figure 6. Optical absorption spectra for 52 & 86 nm silver particulate films deposited on PVP/P4VP blends.

ACKNOWLEDGEMENTS

Pratima Parashar Pandey (Pune) for SEM facility. The author thanks DST for the funding through Women Scientist Scheme (WOS).

REFERENCES

1. Yonzon, C.R., Stuart, D.A., Zhang, X., Mcfarland, A.D., Haynes, C.L. and Duyne, R.P.V. (2005) Towards Advanced Chemical and Biological Nanosensors—An Overview. Talanta, 67, 438-448.

2. Son, W., et al. (2004) Preparation of Antimicrobial Ultrafine Cellulose Acetate Fibers with Silver Nanoparticles. Macromolecular Rapid Communications, 25, 1632-1637.

3. Doty, R.C., et al. (2005) Stabilization and Functionalization of Metallic Nanoparticles. Chemistry of Materials, 17, 4630-4635.

4. Heilmann, A. (2002) Polymer Films with Embedded Metal Nanoparticles. Springler, Berlin.

5. Mayer, A.B.R. (2001) Colloidal Metal Nanoparticles Dispersed in Amphiphilic Polymers. Polymers for Advanced Technologies, 12, 96-106

6. Beecroft, L.L. (1997) Nanocomposite Materials for Optical Applications. Chemistry of Materials, 9, 1302-1317.

7. Caseri, W. (2000) Nanocomposites of Polymers and Metals or Semiconductors: Historical Background and Optical Properties Macromol. Rapid Communications, 21, 705-722.

8. Eilers, H., Biswas, A., Pounds, T.D. and Norton, M.G. (2006) Teflon AF/Ag Nanocomposites with Tailored Optical Properties. Journal of Materials Research, 21, 2168-2171.

9. Wang, Y., Li, Y., Yang, S., Zhang, G., An, D., Wang, C., Yang, Q., Chen, X., Jing, X. and Wei, Y. (2006) A Convenient Route to Polyvinyl Pyrrolidone/Silver Nanocomposite by Electrospinning. Nanotechnology, 17, 3304-3307

10. Bronstein, L.M., Sidorov, S.N. and Valetsky, P.M. (2004) Nanostructured Polymeric Systems as Nanoreactors for Nanoparticle Formation. Russian Chemical Reviews, 73, 501-515

11. Wang, X., Zuo, J., Keil, P. and Grundmeier, G. (2007) Comparing the Growth of PVD Silver Nanoparticles on Ultra Thin Fluorocarbon Plasma Polymer Films and Self-Assembled Fluoroalkyl Silane Monolayers. Nanotechnology, 18, 265303-265313.

12. Kunz, M.S., Shull, K.R. and Kellock, A.J. (1992) Morphologies of Discontinuous Gold Films on Amorphous Polymer Substrates. Journal of Applied Physics, 72, 4458-4460.

13. Rao, K.M., Pattabi, M., Sainkar, S.R., Lobo, A., Kulkarni, S.K., Uchil, J. and Sastry, M.S. (1999) Preparation and Cha- racterization of Silver Particulate Structure Deposited on Softened Poly(4-vinylpyridine) Substrate. Journal of Physics D: Applied Physics, 32, 2327-2336

14. Rao, K.M. and Pattabi, M. (2001) Effect of Polymer-Metal Particle Interaction on the Structure of Particulate Silver Films Formed on Softened Polymer Substrates. Journal of New Materials for Electrochemical Systems, 4, 11-15.

15. Parashar, P. (2012) Electrical Behaviour of Discontinuous Silver Films Deposited on Compatible Polystyrene/Poly(2- vinylpyridine) Composite. Journal of Materials Science: Materials in Electronics, 23, 468-473.

16. Hassell, T., Yoda, S., Howdle, S.M. and Brown, P.D. (2006) Microstructural Characterization of Silver/Polymer Nanocomposites Prepared Using Supercritical Carbon Dioxide. Journal of Physics: Conference Series, 26, 276-279.

17. Parashar, P. (2012) Structural Properties of Silver Particulate Films Deposited on Softened Polymer Blends of Polystyrene/Poly (2-vinylpyridine). Journal of Materials Science: Materials in Electronics, 23, 1169-1173.

18. Parashar, P., Ramakrishna, K., Ramaprasad, A.T. (2011) A Study on Compatibility of Polymer Blends of Polystyrene/Poly(4-vinylpyridine). Journal of Applied Polymer Science, 12, 1729-1735.

19. Haoyan, W. and Eilers, H. (2008) Electrical Conductivity of Thin-Film Composites Containing Silver Nanoparticles Embedded in a Dielectric Fluoropolymer Matrix. Thin Solid Films, 517, 575-581.

20. Kiesow, A., Morris, J.E., Radehaus, C. and Heilmann, A. (2003) Switching Behavior of Plasma Polymer Films Containing Silver Nanoparticles. Journal of Applied Physics, 94, 6988-6990.

21. Parashar, P., Pattabi, M. and Gurumurthy, S.C. (2009) Electrical Behaviour of Discontinuous Silver Films Deposited on Softened Polystyrene and Poly(4-vinylpyridine) Blends. Journal of Materials Science: Materials in Electronics, 20, 1182-1185.

22. Rao, K.M., Pattabi, M., Mayya, K.S., Sainkar, S.R. and Sastry, M.S. (1997) Preparation and Characterization of Silver Particulate Films on Softened Polystyrene Substrates. Thin Solid Films, 310, 97-101.

23. Fritzsche, W., Porwol, H., Wiegand, A., Boronmann, S. and Khler, J.M. (1998) In-Situ Formation of Ag-Containing Nano- particles in Thin Polymer Film. Nanostructured Materials, 10, 89-97.

24. Akamatsu, K., Takei, S., Mizuhata, M., Kajinami, A., Deki, S., Fujii, M., Hayashi, S. and Yamamoto, K. (2000) Preparation and Characterization of Polymer Thin Films Containing Silver and Silver Sulfide Nanoparticles. Thin Solid Films, 359, 55-60.

25. Carotenuto, G. (2001) Synthesis and Characterization of Poly(N-vinylpyrrolidone) Filled by Monodispersed Silver Clusters with Controlled Size. Applied Organometallic Chemistry, 15, 344-351.

26. Kim, J.Y., Shin, D.H., Ihn, K.J. and Suh, K.D. (2003) Amphiphilic Polyurethane-co-Polystyrene Network Films Containing Silver Nanoparticles. Journal of Industrial and Engineering Chemistry, 9, 37-44.

27. Heilmann, A., Quinten, M. and Werner, J. (1998) Optical Response of Thin Plasma-Polymer Films with Non-Spherical Silver Nanoparticles. The European Physical Journal B, 3, 455-461.

28. Takele, H., Greve, H., Pochstein, C., Zaporojtchenko, V. and Faupel, F. (2006) Plasmonic Properties of Ag Nanoclusters in Various Polymer Matrices. Nanotechnology, 17, 3499-3505.

29. Mandal, S., Arumgam, S.K., Pasricha, R. and Sastry, M. (2005) Silver Nanoparticles of Variable Morphology Synthesized in Aqueous Foams as Novel Templates. Bulletin of Materials Science, 28, 503-510.

30. Heilmann, A., Kiesov, A., Gruner, M. and Kreibig, U. (1999) Optical and Electrical Properties of Embedded Silver Nanoparticles at Low Temperatures. Thin Solid Films, 343-344, 175-178.

8

FUNCTIONALIZED CELLULOSE NETWORKS FOR EFFICIENT OIL REMOVAL FROM OIL-WATER EMULSIONS

Uttam C. Paul, Despina Fragouli *, Ilker S. Bayer and Athanassia Athanassiou

Smart Materials, Nanophysics, Istituto Italiano di Tecnologia, via Morego 30, 16163 Genova, Italy

ABSTRACT

The separation of oil from water in emulsions is a great environmental challenge, since oily wastewater is industrially produced. Here, we demonstrate a highly efficient method to separate oil from water in non-stabilized emulsions, using functionalized cellulose fiber networks. This is achieved by the modification of the wetting properties of the fibers, transforming them from oil- and water-absorbing to water-absorbing and oil-proof. In particular, two diverse layers of polymeric coatings, paraffin wax and poly(dimethylsiloxane)-*b*-poly(ethylene oxide) (PDMS-*b*-PEO) diblock copolymer, are applied on the surface of each individual fiber by a two-step dip adsorption process. The resulting cellulose networks exhibit superhydrophilicity and underwater superoleophobicity and they are mechanically reinforced. Therefore, the described treatment makes cellulose fiber networks excellent candidates for the filtration and subsequent removal of oil from oil-in-water non-stabilized emulsions with oil separation efficiency up to 99%. The good selectivity, reproducibility, and cost effectiveness of the preparation process leads to the production of low cost filters that can be used in oil–water separation applications.

KEYWORDS

cellulose; emulsions; oil–water separation; hydrophilicity; underwater superoleophobicity

1. INTRODUCTION

Owing to the large quantities of industrial emulsified wastewater from petrochemical, chemical, and mineral industries, as well as frequent oil spill accidents, oil–water separation is an important ecological problem [1,2]. As a result, a great deal of effort has been dedicated to the surface modification of polymeric woven or non-woven porous materials, turning them into systems able to efficiently separate oil from water. Such materials present advantages due to the large surface-to-volume ratio, but also due to the synergistic effect of the properties of the individual components such as the mechanical properties of the polymeric support and the functionalities of the developed organic or inorganic layers. In particular, hydrophobic-oleophilic or hydrophilic-oleophobic functionalities are introduced on filter paper (FP), polymer films, foams, and textiles [2,3,4,5,6,7,8,9,10,11,12,13,14,15] upon the utilization of macromolecular layers, nano or micro-particles, nanofibers, *etc.*, making these materials able to separate floating oils on water or water–oil emulsions.

Cellulose networks are porous, micro-textured materials [7,16], with a strong affinity for water due to the presence of the abundant hydroxyl groups of cellulose. Numerous efforts have been made in order to turn these materials into water-resistant multifunctional systems, thus expanding their applicability in diverse technological fields [17,18,19,20]. To this end, studies have been conducted on the surface modification of cellulosic FP, which is commonly used for particle separation/filtration processes [9], in order to be utilized as an oil–water separator [5,6,7,8,9,10,11]. However, the majority of the proposed treatments demand complex preparation processes, some of the treatments have a high cost, the materials utilized limit the recyclability of the final system, and in some cases is difficult to scale up the fabrication process [6,7]. Nevertheless, utilization of cellulose fibers in large scale water-oil separation applications presents great advantages due to their abundance, low cost, biodegradability and excellent thermal and chemical stability. Therefore, the ideal cellulose substrate treatment should be of low cost and easily reproduced, the resulting material should be recyclable, and it should have high separation efficiency [6,21].

Herein, we describe a novel surface-treated FP for oil–water separation from non-stabilized emulsions prepared using commercially available hydrophobic paraffin wax (PFW) and Poly(dimethylsiloxane)-*b*-poly(ethylene oxide) (PDMS-*b*-PEO) diblock copolymer. The PFW treatment makes the FP hydrophobic and oleophilic. The subsequent application of PDMS-*b*-PEO offers the final desired properties to the FP, transforming it into a superhydrophilic and underwater superoleophobic porous material, able to separate oil from oil-in-water non-stabilized emulsions efficiently. In contrast to traditional separation processes, such as ultrafiltration membranes, where the applied driven-pressure for emulsion separation is hundreds of kPas, the proposed method demands much less energy in order to separate water from oil effectively. In fact, the separation process is gravity-driven [14,22], and allows to the water phase to pass through the FP, which at the same time completely repels the dispersed oil drops constricting them to remain on its surface, with separation efficiency of 99%. Key filter features including morphology, surface chemistry, wettability and mechanical properties show that PFW/PDMS-*b*-PEO treated FP can be successfully used for the separation of oil from oil-in-water, non-stabilized emulsions.

2. MATERIALS AND METHODS

2.1. Materials

Toluene, methanol, paraffin wax (ASTM D 127, MP 70 °C–80 °C), light white mineral oil (density 0.84 g/mL at 25 °C), methylene blue, methyl orange, sudan blue dyes, and Whatman filter paper (grade 4, diameter 90 mm) were purchased from Sigma Aldrich, St. Louis, MO, USA. Other oils like olive oil, sunflower oil, and corn oil were purchased from a local market. Poly(dimethylsiloxane)-*b*-poly(ethylene oxide) (PDMS-*b*-PEO, M_w: 600 g/mol, PDMS:PEO 25:75) was purchased from Polysciences Inc., Northampton, UK. All the above chemicals were used without any further purification. Deionized water was obtained from Milli-Q Advantage A10 ultrapure water purification system (Millipore, Billerica, MA, USA).

2.2. Fabrication of Superhydrophilic and Underwater Superoleophobic FP

PFW grains (average particle size 2 mm) were dispersed in toluene (5% and 10%, *w*/*v*) and the solutions were heated to 100 °C until clear solutions were obtained. The FP was then immersed in the warm (about 75 °C–80 °C) PFW solution for 30 s, removed and dried under an aspirated laboratory hood for about 1 h first and then in an air oven at 60 °C for 4 h. The PFW treated FPs were then dipped in PDMS-*b*-PEO/methanol solutions (0.5%, 1.0%, 1.5%, and 2.0%, *w*/*v*) for 30 s, removed and dried in the same way as in the PFW treatment. A schematic of the fabrication process and the identification names of the treated FP at different treatment conditions are given in Figure 1 and Table S1 (Supplementary Materials).

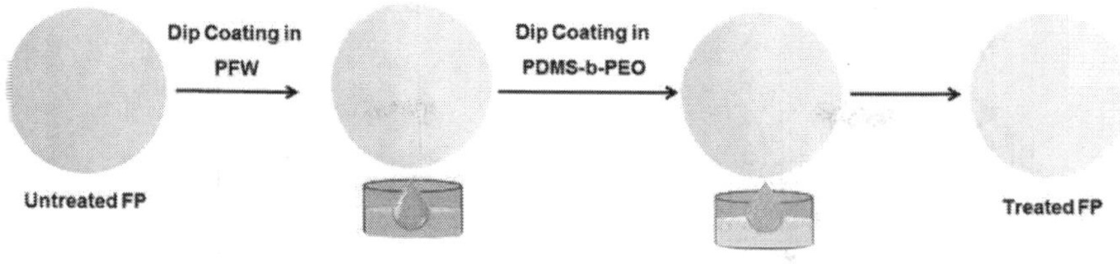

Figure 1. Schematic representation showing the fabrication steps of PFW/PDMS-*b*-PEO-treated FP.

2.3. Characterization Studies

2.3.1. Scanning Electron Microscopy (SEM)

The surface morphology of the FPs was investigated by scanning electron microscopy (SEM), using a variable pressure Jeol JSM-6490LA (JEOL, Tokyo, Japan) microscope equipped with a tungsten (W) thermionic electron source working in high vacuum mode, with an acceleration voltage of 10 kV. The test specimens (0.5 cm × 0.5 cm *c.a.*) were mounted on aluminum stubs with carbon tape. Before starting the measurements the samples were sputter coated with 10 nm of gold using a high resolution sputter coater (Cressington 208 HR, Cressington Scientific Instruments Ltd., Watford, UK).

2.3.2. Porosity Measurements

The FP porosity was measured by completely filling the pores of treated and untreated paper samples of known weight and dimensions with silicon oil (Sigma Aldrich, viscosity 500 Cst, density 0.97 g/mL, and surface tension 21 dynes/cm), as stated in [23]. For this, 90 mm diameter FPs (untreated and treated) were first weighed in a precision balance and then each one was immersed in 10 mL silicon oil for 30 min. After immersion in silicon oil, the FP was gently pressed between two tissue papers in order to remove the excess surface oil. The porosity of the FP was then determined using Equation (1).

$$Porosity\ (\%) = \frac{(C_f - C_i)\,\rho}{C_f} \times 100 \qquad (1)$$

Here, C_i is the weight of FP before silicon oil saturation, C_f is the weight of the silicon oil saturated FP, and ρ is the density of silicon oil.

2.3.3. Fourier Transform Infrared Spectroscopy-Attenuated Total Reflection (FTIR-ATR)

FTIR-ATR spectra of the FPs (treated and untreated) were recorded by Brucker V70 FTIR instrument (Bruker Analytik GmbH, Rheinstetten, Germany) equipped with an ATR unit (zinc selenide–ZnSe crystal). Scanning was run from 4000 to 400 cm^{-1} with 64 repetitive scans averaged for each spectrum and scan resolution was 4 cm^{-1}. Spectra results obtained for the treated FP were compared with that of the untreated FP.

2.3.4. Mechanical Characterization

The mechanical properties of the FPs (treated and untreated) were measured according to ASTM D 638-02a standard test method for tensile properties of plastics, at ambient laboratory conditions (*i.e.*, 21 °C and 50% RH). The dog bone-shaped specimens of the samples were characterized using a 5 kN Instron dual column tabletop universal testing system with 0.5 kN preload (T.A. Instruments, Instron model 3365, Norwood, MA, USA) and a crosshead rate of 5 mm/min. The tensile strength, percentage elongation at break, and Young's modulus were calculated for each specimen. At least five samples were tested for each type of treatment and the results were averaged to obtain a mean value.

2.3.5. Wettability and Contact Angle Measurements

Static and dynamic water contact angle measurements were carried out on rectangular strip samples using a contact angle goniometer DataPhysics OCAH 200 (Kruss GmbH, Hamburg, Germany) equipped with a CCD camera and image processing software, operating under laboratory conditions. Water droplets of volume 5 and 2 μL were placed on the surface of the samples and the static contact angle was measured within 45 s. The dynamic contact angle was recorded in a time range between 0.0 s (immediately after the droplet deposition) and 3.5 s. Up to ten measurements were conducted at random locations for each treated sample and the results were averaged to obtain a mean value. For underwater static and dynamic oil contact angle testing, the untreated and treated FPs were firmly placed at the bottom of a square optical glass cell (GC 10, no. 6000017, Data Physics, Kruss GmbH, Hamburg, Germany) with double-sided tape. The glass cell was filled with 6 mL of millipore water in order to completely cover the sample. A 12 μL dichloromethane (DCM) droplet was generated underwater on the paper sample and the contact angle was measured. At least 5 measurements at random positions were conducted and an average value was calculated.

2.3.6. Separation of Oil from Oil-in-Water Non-Stabilized Emulsions

Separation of oil and water was carried out as per the setup illustrated in the Figure 2. Two types of oil-in-water emulsions were studied: the first consisting of 80% water and 20% oil, (0.01 wt % of methylene blue 59 dye was used for water coloring, for better visualization and distinction of the two liquids), and the second consisting of 90% water and 10% oil. The oil-in-water emulsion was prepared with a vortex mixer (Heidolph multi reax, Schwabach, Germany) at 500 rpm for 10 min and then the emulsion was poured onto the FP through a glass tube. It must be noted that no surfactants or emulsion stabilizers were used during sonic treatment of the oil–water mixtures. The resultant non-stabilized emulsions separation process during the gravity driven filtration process is depicted in Figure 2.

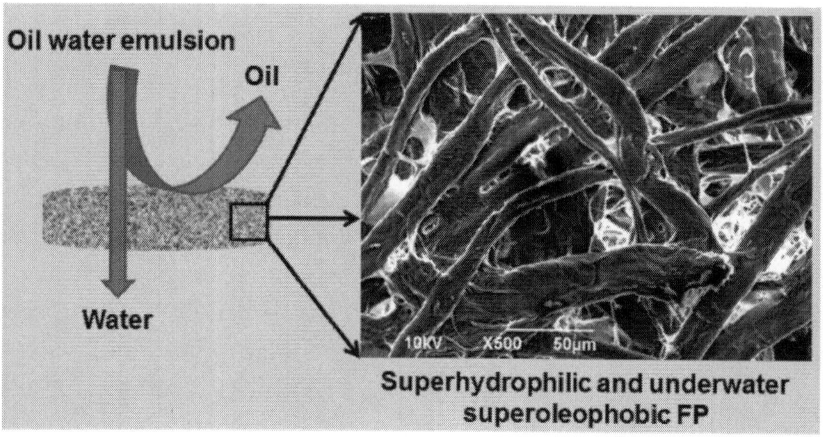

Figure 2. Schematic representation of the separation of oil and water from a non-stabilized oil-in-water emulsion.

In all cases, the separation process is gravity driven. After the separation processes, the oil and water recovery (%) are determined to be the volume ratio of the separated over the initial oil or water, respectively, as shown in Equations (2) and (3).

$$Water\ Recovery\ (\%) = \frac{V_{collected\ water}}{V_{initial\ water}} \times 100 \quad (2)$$

$$Oil\ Recovery\ (\%) = \frac{V_{collected\ oil}}{V_{initial\ oil}} \times 100 \quad (3)$$

From Table S2 (Supplementary Materials), it can be seen that according to the preliminary results for a 10 mL emulsion (20% of oil-in-water), the treatment of the FP with 10% *w/v* PFW and 1.0% *w/v* PDMS-*b*-PEO in methanol exhibited better separation performance. Based on this, the untreated, 10% PFW, 1.0% PDMS-*b*-PEO and 10% PFW and subsequently 1.0% PDMS-*b*-PEO treated FP results are presented in this manuscript.

2.3.7. UV-vis Measurements

In the separation system, the concentration of the mineral oil after filtration and the separation efficiency were estimated using a Varian CARY 300 Scan UV-visible spectrophotometer (Agilent Technologies, Santa Clara, CA, USA) according to a standard procedure [8,24]. Initially the absorption spectra of 2 mL of different concentrations (0.0–30.0 mg/L) of oil solutions were recorded after diluting the mineral oil in hexane. The intensity of the absorbance peak at 272 nm for each oil concentration was then plotted as shown in Figure S1 (Supplementary Materials) and the equation which gives the dependence of the intensity to the oil concentration was defined by a linear fit. For these measurements the baseline correction of hexane was used. Subsequently the absorption spectra of 2 mL of filtered oil, water and original (pristine) oil were recorded, and the concentration of oil was calculated by the intensity of the peak and the above mentioned linear fit inFigure S1 (Supplementary Materials). At least three repetitions were made for the determination of oil concentration in each sample.

3. RESULTS AND DISCUSSION

Two diverse polymeric layers were applied on the FP, namely PFW and PDMS-*b*-PEO, by a two-step dip adsorption process forming a functional coating on the surface of each individual fiber. PFW is a white, odorless, water-insoluble long chain branch hydrocarbon molecule which contains 20 to 40 carbon atoms in its chain, and is used as a coating, water/moisture barrier, anti-caking and lubricating agent for paper and textiles [25,26,27,28]. PDMS-*b*-PEO is a nontoxic environmentally friendly hydrophilic semi crystalline diblock copolymer [29,30,31,32], which is used in order to modify the surface properties of polymers, e.g., transform a hydrophobic material into a hydrophilic one [33,34].

Figure 3 shows the scanning electron microscope images (SEM) of the original (untreated) and the modified FP coated with PFW, PDMS-*b*-PEO, and PFW/PDMS-*b*-PEO. The untreated paper shows the interconnecting network of cellulose fibers with diameters of a few tens of micrometers, as shown in Figure 3a. When the untreated FP is dipped in the PFW solution, the latter penetrates in the innermost volume of the FP and coats all the fibers, forming a layer of 1.0 ± 0.1 μm as the cross section TEM analysis revealed (Supplementary Materials, Figure S2a,b). This thin layer does not significantly affect the network porosity and the morphology of the fibers which are very similar before and after the treatment as proved by the SEM micrograph in Figure 3b. After PFW treatment, FPs were subsequently treated with the PDMS-*b*-PEO solutions by dip-coating, and the overall morphology and structure of the porous network remained practically unaltered (Figure 3c,d). In fact, according to the cross section TEM analysis (Supplementary Materials, Figure S2c) the thickness of the polymer coating remains the same, 1.0 ± 0.1μm. The use of higher PFW concentrations did not show the same effect on the coating of the fibers since the polymer was forming a homogeneous compact film on the surface of the FP. On the other hand, higher PDMS-*b*-PEO quantities did not change the morphology of the treated paper.

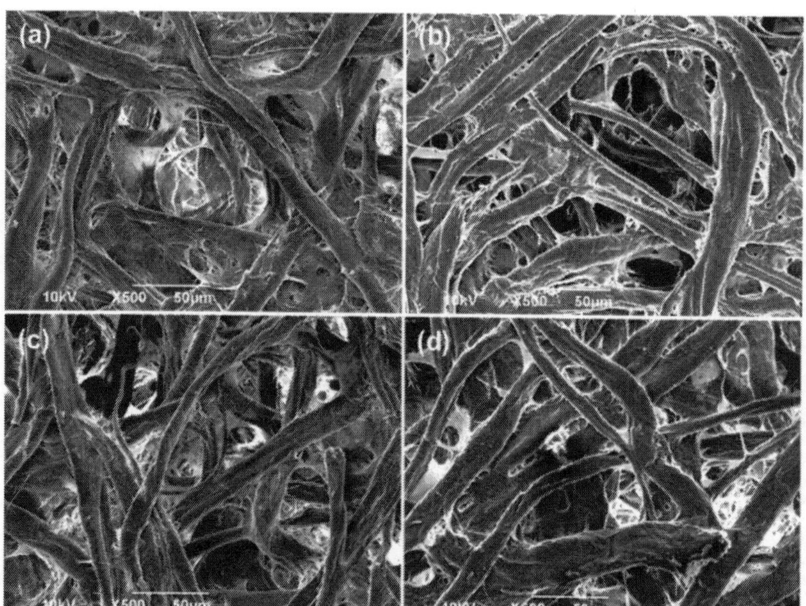

Figure 3. Scanning electron microscopy (SEM) surface analysis of (**a**) untreated, (**b**) PFW treated, (**c**) PDMS-*b*-PEO treated, and (**d**) PFW /PDMS-*b*-PEO treated FP samples.

The porosity study of the untreated and treated FPs shows that the applied PFW layer causes a slight decrease in the porosity of the untreated sample. In fact, as shown in the Supplementary Materials (Table S3), the mean porosity of the FP (53%), is decreased to 43% after the PFW treatment. The subsequent application of the PDMS-*b*-PEO layer does not significantly alter the porosity of the PFW coated FP (41% for the PFW/PDMS-*b*-PEO). Similarly, the sole PDMS-*b*-PEO treatment of the cellulose fibers does not have significant effect on the porosity (53.5%) of the untreated sample. This indicates that the PFW treatment forms a thick solid layer on the surface of each individual fiber whereas the PDMS-*b*-PEO layer is much thinner.

Figure 4 shows the FTIR-ATR spectra of untreated and treated FPs. Typical FTIR-ATR spectra of cellulose show a broad, strong absorption peak at about 3331 cm^{-1} due to the O–H stretching vibrations arising from abundant free hydroxyl and hydrogen-bonded hydroxyl groups, and the O–H bending of the adsorbed water at 1632 cm^{-1}. The vibrations at 2918, 2851, and 1310 cm^{-1} are attributed to C–H stretching, bending and rocking, respectively. The absorption band at 1427 cm^{-1} is assigned to the C–H scissoring motion of cellulose. Finally, the signals of anti-symmetrical bridge C–O stretching and C–O–C stretching vibrations appear at 1157 and 1030 cm^{-1} of cellulose [35, 36].

In case of PFW and PFW/PDMS-*b*-PEO treated filters all the absorption bands are present almost in the same position as in the spectra of the pristine cellulose FP. As shown in Figure 4, the strong and broad band around 3331 cm^{-1} that corresponds to the –OH stretching vibrations of hydroxyl groups, is practically at the same position for the untreated, PFW, and PFW/PDMS-*b*-PEO -treated cellulose. The characteristic absorption bands of carbon chains (C–H) for cellulose are at 2851 and 2918 cm^{-1}. These bands are also at the same position in case of PFW and PFW/PDMS-*b*-PEO treatment. Therefore, there are no specific differences between treated and untreated FPs indicating that the PFW and PDMS-*b*-PEO layers are bonded to the cellulosic FP only through strong physical interactions.

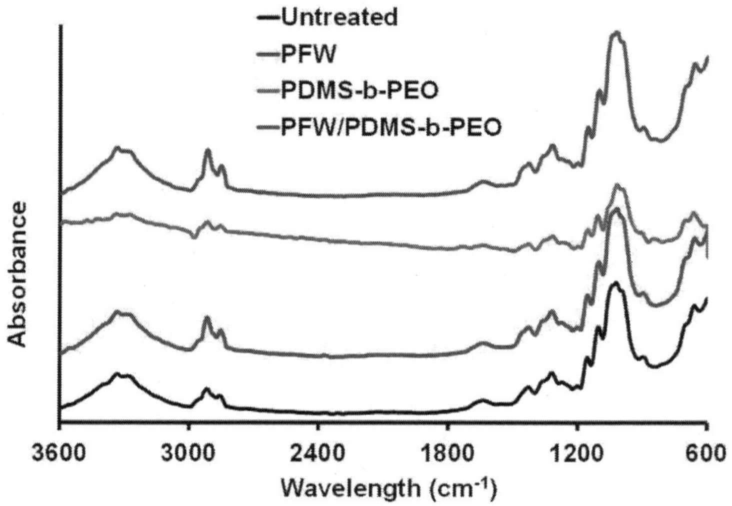

Figure 4. FTIR-ATR spectra of untreated FP, PFW, PDMS-*b*-PEO, and PFW/PDMS-*b*-PEO treated FP samples.

The mechanical strength of the pristine FP, PFW, PDMS-*b*-PEO, and PFW/PDMS-*b*-PEO -treated FPs was studied, and the results of Young's modulus, tensile strength, and percentage elongation at break are shown in Figure 5. The elastic modulus of the FPs after being treated with PFW increases from 521.3 to 1.2 GPa, while their tensile strength increases from 3.8 to 6.1 N (Figure 5b). At the same time, the elongation at break decreases from 9.4% to 4.1%. We assume that the strong adhesion of the PFW on the cellulose fibers can induce physical cross-linking of the fibers and PFW, which turns the FP into a stronger material [17]. On the other hand, the PDMS-*b*-PEO treatment does not seem to have any influence to the mechanical properties of the FPs both in the untreated and PFW treated samples. In fact, after treatment of the PFW treated filters with the PDMS-*b*-PEO polymer, the excellent mechanical properties obtained due to the PFW are still maintained.

Water and oil (dichloromethane–DCM) static and dynamic contact angle tests in air and underwater on the untreated, PFW, PDMS-*b*-PEO, and PFW/PDMS-*b*-PEO treated FPs were carried out to investigate their wettability properties (Table 1,Figure 6 and Figure S3 Supplementary

Materials). The untreated FP is rapidly wetted by water (after 80 ms) while the oil droplets are absorbed in less than few milliseconds (time below the instrument's resolution), due to the abundant hydroxyl groups and cavities in its structure. In case of the PFW treatment, a hydrophobic coating is formed around each cellulose fiber, preventing water from wetting and penetrating. As shown in Table 1, Figure 6a and in Supplementary Materials Figures S3 and S4, when a water droplet is placed on the PFW-treated FP surface, it steadily remains on its surface, forming a water contact angle of 125.8° ± 2.3°. Conversely, the oil droplet wets the substrate immediately (faster than the instrument's resolution). It should be noticed that lower amounts of PFW for the FP treatment were not sufficient to homogeneously coat the cellulose fibers, and therefore the water was slowly absorbed by the paper.

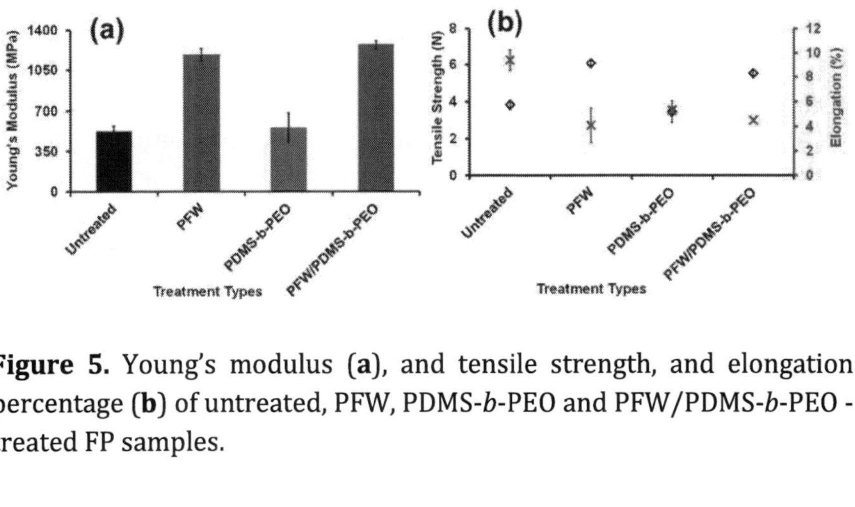

Figure 5. Young's modulus (**a**), and tensile strength, and elongation percentage (**b**) of untreated, PFW, PDMS-*b*-PEO and PFW/PDMS-*b*-PEO - treated FP samples.

The subsequent PDMS-*b*-PEO treatment of the FP/PFW sample causes a remarkable change in the wetting properties, transforming the treated FP into a hydrophilic/oleophilic material with the water droplet being completely absorbed after 3.5 s, while the oil is absorbed immediately after the deposition (Figures S3 and S4 Supplementary Materials). In particular after dipping the PFW-treated FP into the PDMS-*b*-PEO solution, the PDMS block interacts with the PFW layer due to van der Waals forces and hydrophobic interactions, while the PEO, as a hydrophilic block, is exposed to the surface [31,32]. Therefore, the FP's surface composition and properties depend on the fraction of the exposed PEO hydrophilic component. Higher amounts of PDMS-*b*-PEO for the treatment did not show any further improvement, while lower amounts were not enough so as to coat all the fibers and therefore the hydrophilicity of the FPs was reduced. The effectiveness of the PDMS-*b*-PEO component to hydrophilize surfaces is also demonstrated for the pure FPs, with the paper becoming even more hydrophilic and thus increasing the absorption speed from 80 to 60 ms, possibly due to the interaction of the copolymer with the hydrophobic components of the surface of the cellulose fibers. In the case of underwater oil contact angles, untreated, PFW, and only PDMS-*b*-PEO treated FPs are oleophobic with a stable oil contact angle over time of 134.7° ± 3.2°, 140.0° ± 1.6°, 142.4° ± 0.1° respectively. Surprisingly, when an oil droplet is placed on the surface of the PFW/PDMS-*b*-PEO FP underwater, it forms a static contact angle of 148.7° ± 6.40°, Table 1, and Figure 6d. In fact, due to the hydrophilicity of the treated FP, a water cushion between the oil droplet and the cellulose surface is formed underwater, with trapped water molecules within the FP network. This effect establishes the necessary conditions for the underwater superoleophobicity due to the repulsive interaction between polar (water) and non-polar (oil) molecules and the differences of their surface tension [1,14,15]. Underwater oleophobic properties are also observed when the oil droplet is placed on the wet PFW/PDMS-*b*-PEO FP surface in air as shown in Figure S4 (Supplementary Materials). On this wetted FP the oil droplet was not absorbed. Conversely, when the same experiment is repeated on the untreated, PFW and PDMS-*b*-PEO treated wet surface in air, the oil is absorbed within 1 min (see Figure S5 Supplementary Materials).

The underwater superoleophobic properties of the PFW/PDMS-*b*-PEO FP make it a promising candidate for oil–water separation applications. As a proof of concept, the oil-in-water emulsion separation is performed and sequentially shown inFigure 7. Ten milliliters of non-stabilized oil-in-water emulsion (80% water, and 20% oil) was poured onto the differently treated FPs. The untreated, PFW and PDMS-*b*-PEO -treated FPs were not able to completely separate the oil from the emulsion, as oil was completely adsorbed on the surface of the samples and its traces were observed on the filtered water (Figure S6, Supplementary Materials). However, due to the superhydrophilic and the underwater superoleophobic nature of the PFW/PDMS-*b*-PEO-treated FP, water can easily wet and penetrate the treated sample while the oil is trapped on its surface. As the microscopy study of the emulsion before and after the separation process showed, the maximum size of the non-stabilized emulsion droplets is 60 μm, (see Supplementary Materials Figure S7) while the minimum droplet size that can be separated by this process is 1.2 μm.

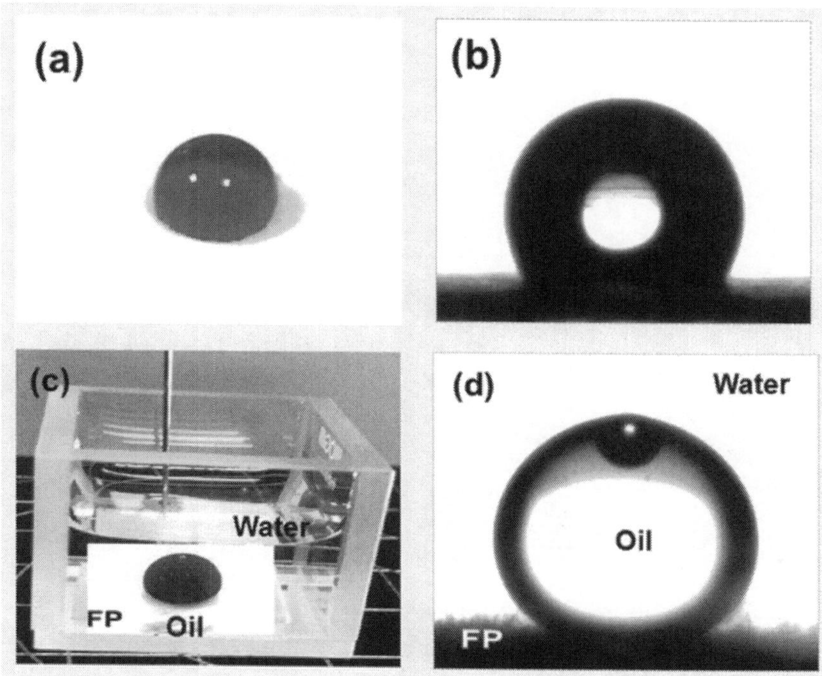

Figure 6. (**a**) Water droplet on PFW-treated FP. Methyl orange dye (0.01 wt %) is used to color the water, (**b**) water contact angle (125.8° ± 2.3°) of PFW-treated FP, (**c**) underwater DCM droplet on PFW/PDMS-*b*-PEO-treated FPs. Sudan blue dye (0.01%) is used to color the DCM, and (**d**) underwater DCM contact angle (148.7° ± 6.4°) of PFW/PDMS-*b*-PEO-treated FPs.

Table 1. Water and oil (DCM) static contact angle (SCA) measurement of untreated, PFW, PDMS-*b*-PEO and PFW/PDMS-*b*-PEO treated FP in air and underwater.

Sample	SCA of water (°) air	SCA of DCM (°) air	SCA of DCM (°) underwater
Untreated	absorbed	absorbed	134.7 ± 3.2
PFW	125.8 ± 2.3	absorbed	140.0 ± 1.6
PDMS-*b*-PEO	absorbed	absorbed	142.4 ± 0.6
PFW/PDMS-*b*-PEO	absorbed	absorbed	148.7 ± 6.4

Figure 7. Different steps of oil–water separation experiments on 10 mL of oil-in-water non-stabilized emulsion. (**a**) Oil/water emulsion. Inset shows zoomed view for oil/water emulsion. Here, water is colored with methylene blue dye, in order to be distinguished from oil; (**b**,**c**) Separated water finishes at the bottom of the flask and separated oil remains on the filter paper; (**d**,**e**) Water and oil stored separately after filtration.

Four different kinds of oils were successfully separated from their water emulsions through the same process. The oils that were used were mineral oil (density: 0.840 g/L), corn oil (density: 0.930 g/L), olive oil (density: 0.860 g/L), and sunflower oil (density: 0.920 g/L). No visible residual oil was observed in the collected water, even after filtration of 100 mL of oil-in-water emulsions. This indicates that the treated FP can separate different kinds of oil with different densities. Since the average filtration time of 10 mL oil-in-water emulsion is about 55 s, then *c.a.* separation rate of oil–water is 77 $L\cdot m^{-2}\cdot d^{-1}$ (77 liters per meter square of FP per day), with the driving force being only the weight of the 10mL emulsion. Emulsions with higher volumes e.g., 1000 mL were also tested and similar results were found in terms of separation efficiency and rate, indicating that the treated FP can effectively separate large volumes of emulsions (Figure S8, Supplementary Materials).

To confirm the high separation efficiency, the potential presence of residual oil content in the filtered water was measured with UV-visible spectroscopy (Figure 8) [8]. As shown, the absorption curve of the filtered water is the same as the absorption curve for reference clean water, proving that there are hardly any oil contaminants present. On the contrary, a characteristic absorption band is observed in the filtered oil (mineral oil) absorption curve, with a peak at 272 nm [37,38] typical for the mineral oil. The absorption intensity of the peak is used in order to define the mineral oil concentration as discussed in the experimental section, and subsequently for the determination of the filtration efficiency of oil using Equation (4):

$$Filtration\ Efficiency = (1 - \frac{C_f}{C_i}) \times 100\% \tag{4}$$

where C_f is the concentration of oil in the filtered water and C_i is the initial concentration of oil in the non-stabilized oil–water emulsion. The filtration efficiency was calculated as 99.0% in accordance with the value calculated by the volumetric ratio.

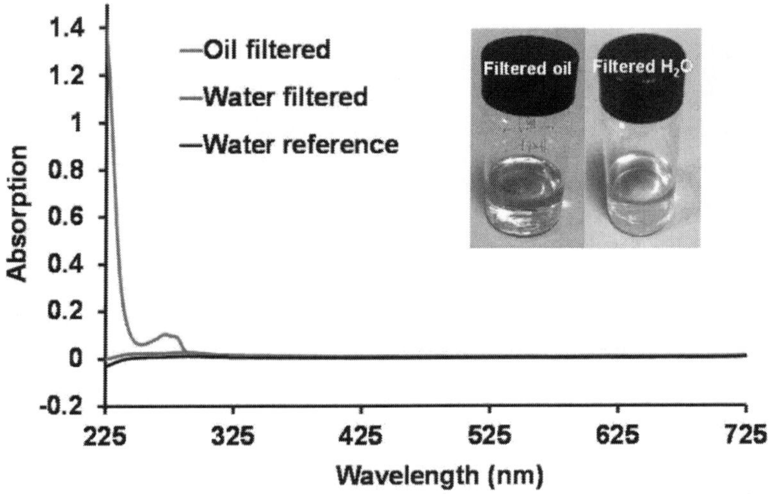

Figure 8. UV-vis absorption spectra of reference water (miliQ), filtered water and filtered oil.

4. CONCLUSIONS

In conclusion, we present a novel method to modify the surface properties of filter paper using polymeric solutions, rendering them superhydrophilic and underwater superoleophobic in a simple, straightforward and cost effective way. Various oils such as mineral oil, olive oil, and vegetable oil can be efficiently separated from their water emulsions based on the gravity-driven separation process. The filter paper becomes superhydrophilic and also exhibits underwater superoleophobic properties after PFW/PDMS-*b*-PEO treatment. The SEM and TEM analysis revealed that the morphology of the filter paper is not altered significantly while the mechanical properties of the system are improved. The superhydrophilic and underwater superoleophobic filter paper possesses excellent oil–water separation efficiency up to 99% and a rate of about 77 $L \cdot m^{-2} \cdot d^{-1}$. As such, the treatment presented here can be easily industrialized due to fact that both the substrate (cellulose) and polymers used are abundant and produced at industrial scales.

ACKNOWLEDGMENTS

The authors acknowledge Tiziano Catelani, Nanochemistry, Istituto Italiano di Tecnologia for performing the TEM analysis.

AUTHOR CONTRIBUTIONS

Uttam C. Paul performed the experiments and manuscript writing; Despina Fragouli, Ilker S. Bayer, and Athanassia Athanassiou carried out project development and leadership, manuscript interpretation and corrections.

CONFLICTS OF INTEREST

The authors declare no conflict of interest.

REFERENCES

1. Wen, Q.; Di, J.; Jiang, L.; Yu, J.; Xu, R. Zeolite-coated mesh film for efficient oil–water separation. *Chem. Sci.* **2013**, *4*, 591–595.

2. Tao, S.; Wang, Y. Synthesis of hierarchically porous silica film with controllable surface wettability. *Int. Nano Lett.* **2014**,*4*, 1–6.

3. Calcagnile, P.; Fragouli, D.; Bayer, I.S.; Anyfantis, G.C.; Martiradonna, L.; Cozzoli, P.D.; Cingolani, R.; Athanassiou, A. Magnetically driven floating foams for the removal of oil contaminants from water. *ACS Nano* **2012**, *6*, 5413–5419.

4. Chu, Z.; Feng, Y.; Seeger, S. Oil/water separation with selective superantiwetting/superwetting surface materials.*Angew. Chem. Int. Ed.* **2015**, *54*, 2328–2338.

5. Yu, T.; Zu, G.; Wang, X.; Yang, J.; Hu, J. Fabrication of oil–water separation filter paper by simple impregnation with fluorinated poly-acrylate emulsion. *BioResources* **2014**, *9*, 4421–4429.

6. Wang, S.; Li, M.; Lu, Q. Filter paper with selective absorption and separation of liquids that differ in surface tension.*ACS Appl. Mater. Interfaces* **2010**, *2*, 677–683.

7. Zhang, M.; Wang, C.; Wang, S.; Shi, Y.; Li, J. Fabrication of coral-like superhydrophobic coating on filter paper for water–oil separation. *Appl. Surf. Sci.* **2012**, *261*, 764–769.

8. Rohrbach, K.; Li, Y.; Zhu, H.; Liu, Z.; Dai, J.; Andreasen, J.; Hu, L. A cellulose based hydrophilic, oleophobic hydrated filter for water/oil separation. *Chem. Commun.* **2014**, *50*, 13296–13299.

9. Kong, L.; Wang, Q.; Xiong, S.; Wang, Y. Turning low-cost filter papers to highly efficient membranes for oil/water separation by atomic-layer-deposition-enabled hydrophobization. *Ind. Eng. Chem. Res.* **2014**, *53*, 16516–16522.

10. Du, C.; Wang, J.; Chen, Z.; Chen, D. Durable superhydrophobic and superoleophilic filter paper for oil–water separation prepared by a colloidal deposition method. *Appl. Surface Sci.* **2014**, *313*, 304–310.

11. Feng, X.; Shi, Y.; Liu, J.; Yang, W. Fabrication of filter paper with tunable wettability and its application in oil–water separation. *J. Sol-Gel Sci. Technol.* **2015**, *4*, 3759–3767.

12. Wang, B.; Liang, W.; Guo, Z.; Liu, W. Biomimetic super-lyophobic and super-lyophilic materials applied for oil/water separation: A new strategy beyond nature. *Chem. Soc. Rev.* **2015**, *44*, 336–361.

13. Xue, C.H.; Ji, P.T.; Zhang, P.; Lia, Y.R.; Ji, S.T. Fabrication of superhydrophobic and superoleophilic textiles for oil–water separation. *Appl. Surface Sci.* **2013**, *284*, 464–471.

14. Liu, X.; Ge, L.; Li, W.; Wang, X.; Li, F. Layered double hydroxide functionalized textile for effective oil/water separation and

selective oil adsorption. *ACS Appl. Mater. Interfaces* **2015**, *7*, 791–800.

15. Seddighi, M.; Hejazi, S.M. Water–oil separation performance of technical textiles used for marine pollution disasters.*Marine Pollut. Bull.* **2015**, *96*, 286–293.

16. Zhang, F.; Zhang, B.W.; Shi, Z.; Wang, D.; Jin, J.; Jiang, L. Nanowire-haired inorganic membranes with superhydrophilicity and underwater ultralow adhesive superoleophobicity for high-efficiency oil/water separation.*Adv. Mater.* **2013**, *25*, 4192–4198.

17. Bayer, I.S.; Fragouli, D.; Attanasio, A.; Sorce, B.; Bertoni, G.; Brescia, R.; Corato, R.D.; Pellegrino, T.; Kalyva, M.; Sabella, S.; *et al.* Water-repellent cellulose fiber networks with multifunctional properties. *ACS Appl. Mater. Interfaces***2011**, *3*, 4024–4031.

18. Ayadi, F.; Bayer, I.S.; Fragouli, D.; Liakos, I.; Cingolani, R.; Athanassiou, A. Mechanical reinforcement and water repellency induced to cellulose sheets by a polymer treatment. *Cellulose* **2013**, *20*, 1501–1509.

19. Jang, I.; Song, S. Facile and precise flow control for a paper-based microfluidic device through varying paper permeability. *Lab Chip* **2015**, *15*, 3405–3412.

20. Zhao, W.; van Den Berg, A. Focus Lab on paper. *Lab Chip* **2008**, *8*, 1988–1991.

21. Liu, Q.; Patel, A.A.; Liu, L. Superhydrophilic and underwater superoleophobic poly(sulfobetaine methacrylate)-grafted glass fiber filters for oil–water separation. *ACS Appl. Mater. Interfaces* **2014**, *6*, 8996–9003.

22. Si, Y.; Fu, Q.; Wang, X.; Zhu, J.; Yu, J.; Sun, G.; Ding, B. Superelastic and superhydrophobic nanofiber-assembled cellular aerogels for effective separation of oil/water emulsions. *ACS Nano* **2015**, *9*, 3791–3799.

23. Ahmed, F.E.; Lalia, B.S.; Hilal, N.; Hashaikeh, R. Underwater superoleophobic cellulose/electrospun PVDF–HFP membranes for efficient oil/water separation. *Desalination* **2014**, *344*, 48–54.

24. Chakrabarty, B.; Ghoshal, A.K.; Purkait, M.K. Ultrafiltration of stable oil-in-water emulsion by polysulfone membrane. *J. Membr. Sci.* **2008**, *325*, 427–437.

25. Spence, K.L.; Richard, A.; Venditti, R.A.; Rojas, O.J.; Pawlak, J.J.; Hubbe, M.A. Water vapor barrier properties of coated and filled microfibrillated cellulose composite films. *BioResources* **2011**, *6*, 4370–4388.

26. Kumar, S.; Kumar, A. Study on petroleum derived waxes and their uses. *Int. J. Res. Sci. Technol.* **2013**, *2*, 1–8.

27. Khwaldia, K. Water vapor barrier and mechanical properties of paper-sodium caseinate and paper-sodium caseinate-paraffin wax films. *J. Food Biochem.* **2010**, *34*, 998–1013.

28. Donhowe, G.; Fennema, O.R. Water vapor and oxygen permeability of wax films. *J. Am. Oil Chem. Soc.* **1993**, *70*, 867–873.

29. Rheingans, O.; Hugenberg, N.; Harris, J.R.; Fischer, K.; Maskos, M. Nanoparticles built of cross-linked heterotelechelic, amphiphilic poly(dimethylsiloxane)-*b*-poly(ethylene oxide) diblock copolymers. *Macromolecules* **2000**,*33*, 4780–4790.

30. Liu, Y.; Zhang, J.; Tu, M. Synthesis of porous methylphenylsiloxane/poly(dimethylsiloxane) composite using poly(dimethylsiloxane)–poly(ethylene oxide) (PDMS–PEO) as template. *J. Mater. Sci.* **2012**, *47*, 3350–3353.

31. Galin, M.; Mathis, A. Structural and thermodynamic study of dimethylsiloxane-ethylene oxide PDMS-PEO-PDMS triblock copolymers. *Macromolecules* **1981**, *14*, 677–683.

32. Luzinov, I.; Minko, S.; Tsukruk, V.V. Adaptive and responsive surfaces through controlled reorganization of interfacial polymer layers. *Prog. Polym. Sci.* **2004**, *29*, 635–698.

33. Yao, M.; Fang, J. Hydrophilic PEO-PDMS for microfluidic applications. *Micromech. Microeng.* **2012**, *22*, 1–6.

34. Chavan, A.A.; Li, H.; Scarpellini, A.; Marras, S.; Manna, L.; Athanassiou, A.; Fragouli, D. Elastomeric nanocomposite foams for the removal of heavy metal ions from water. *ACS Appl. Mater. Interfaces* **2015**, *7*, 14778–14784.

35. Adela, A.M.; El-Wahab, Z.H.A.; Ibrahim, A.A.; Al-Shemya, M.T. Characterization of microcrystalline cellulose prepared from lignocellulosic materials. Part II: Physicochemical properties. *Carbohydr. Polym.* **2011**, *83*, 676–687.

36. Cunha, A.G.; Freire, C.; Silvestre, A.; Neto, C.P.; Gandini, A.; Belgacem, M.N.; Chaussy, D.; Beneventi, D. Preparation of highly hydrophobic and lipophobic cellulose fibers by a straightforward gas–solid reaction. *J. Colloid Interface Sci.* **2010**, *344*, 588–595.

37. Gomez, N.A.; Abonia, R.; Hector Cadavid, H.; Vargas, I.H. Chemical and spectroscopic characterization of a vegetable oil used as dielectric coolant in distribution transformers. *J. Braz. Chem. Soc.* **2011**, *22*, 2292–2303.

38. ASTM D 2269. *Standard Test Method for Evaluation of White Mineral Oils by Ultraviolet Absorption*; ASTM International: West Conshohocken, PA, USA, 1988.

9

COHESIVE STRENGTH OF NANOCRYSTALLINE ZNO:GA THIN FILMS DEPOSITED AT ROOM TEMPERATURE

Anura Priyajith Samantilleke, Luís Manuel Fernandes Rebouta Vitor Garim, Luis Rubio-Peña, Senetxu Lanceros-Mendez, Pedro Alpuim, Sandra Carvalho, Alexey V Kudrin and Yury A Danilov

1Centro de Física, Universidade do Minho, Azurém, 4800-058 Guimarães, Portugal

2Engineering School, University of Cadiz, C/Chile, 1. 11002 Cádiz, Spain

3Physical-Technical Research Institute, N. I. Lobachevskiy State University, Nihzniy Novgorod, Russia

ABSTRACT

In this study, transparent conducting nanocrystalline ZnO:Ga (GZO) films were deposited by dc magnetron sputtering at room temperature on polymers (and glass for comparison). Electrical resistivities of 8.8×10^{-4} and 2.2×10^{-3} Ω cm were obtained for films deposited on glass and polymers, respectively. The crack onset strain (COS) and the cohesive strength of the coatings were investigated by means of tensile testing. The COS is similar for different GZO coatings and occurs for nominal strains approx. 1%. The cohesive strength of coatings, which was evaluated from the initial part of the crack density evolution, was found to be between 1.3 and 1.4 GPa. For these calculations, a Young's modulus of 112 GPa was used, evaluated by nanoindentation.

1. INTRODUCTION

Doped ZnO thin films are widely used as transparent electrodes in optoelectronic and electro-optic devices such as solar cells and flat panel displays [1–3], because of their unique properties, specifically low electrical resistivity and high transmittance in the visible spectral region [4]. These properties are obtained using substrate temperatures higher than 200°C, but growing interest in flexible substrates has led to the use of polymeric alternatives, which require the deposition of films at low temperature [5]. Furthermore, the deposition on polymeric substrates decreases the quality of the film properties [6]; therefore, the pursuit toward an understanding of the structural, electromechanical and electro-optical properties of nanocrystalline (nc) thin films is crucial for device applications.

2. EXPERIMENTAL DETAILS

ZnO:Ga (GZO) thin films were deposited by dc-magnetron sputtering on glass and polyethylene naphthalate (PEN) substrates, under an Ar atmosphere with a base pressure of 2×10^{-4} Pa, from a GZO target (zinc oxide/gallium oxide, 95.5/4.5 wt.%) of 2" diameter. A target current density of 0.6 mA/cm^2 was applied, and a deposition rate of 21 nm/min was obtained. No bias was applied to the substrate holder during the depositions, which took place at room temperature. The working pressure (P_w) was varied from 0.41 to 0.86 Pa, with the target-to-substrate distance kept at a constant 8 cm. The crystallinity and crystal orientation was studied using a Bruker AXS Discover D8 (Madison, USA) for X-ray diffraction (XRD). Glass substrates were used to avoid the presence of polymer substrate peaks. The electrical resistivity, carrier concentration and Hall mobility of the coatings on glass substrates were all measured using Van der Pauw geometry under a magnetic field of 1

Tesla. The electromechanical tests were carried out on 10 × 40 mm^2 samples using a computer-controlled tensile testing machine (Minimat, Polymer Labs, Loughborough, UK), which was mounted on an optical microscope stage (Nikon Optiphot-100, Tokyo, Japan). One of the grips of the instrument was displaced at a constant speed of 0.2 mm/min. The applied load and stage displacement values were recorded at 1-s intervals. Crack development was recorded through a CCD camera connected to the microscope, with the evolution of the crack density obtained by the subsequent video analysis. The thickness of the polymer substrates was measured using a Fischer Dualscope MP0R instrument (Sindelfingen, Germany).

3. RESULTS AND DISCUSSION

3.1 Structural characterization

Figure 1a shows the XRD spectra obtained for nc GZO thin films (approx. 100-nm thick) as a function of the P_w, where only the ZnO (002) peak, at approx. 34°, is observed. The spectra reveal a highly textured hexagonal phase with a wurtzite structure. A lower P_w resulted in samples with a higher c-lattice parameter. In the thin films prepared with a P_w, between 0.41 and 0.86 Pa, the (002) peak position shifted from 2θ = 33.93° (c = 0.528 nm) to 2θ = 34.06° (c = 0.525 nm). The full-width at half-maximum (FWHM) can be expressed as a linear combination of the lattice strain and crystalline size. The effects of strain and particle size on the FWHM can be expressed as [7]

$$\beta\cos\theta/\lambda = (1/\varepsilon) + (\tau\sin\theta/\lambda) \tag{1}$$

where β is the measured FWHM, θ is the Bragg angle of the peak, λ is the X-ray wavelength (1.5418 Å), ε is the effective particle size and τ is the effective strain. The average particle size, calculated from the plot cos θ versus sin θ shown in Figure 1b, was 8.7 nm. The particle size (D_v) calculated from Scherrer's formula ($D_v = 0.94\lambda/(\beta\cos\theta)$), was 8.9 nm, which is very close to that calculated from Equation 1 [8]). The presence of strain in the ZnO crystal lattice, caused indirectly by P_w, can be expected to exert significant influence on the mechanical properties of the nc-GZO thin film.

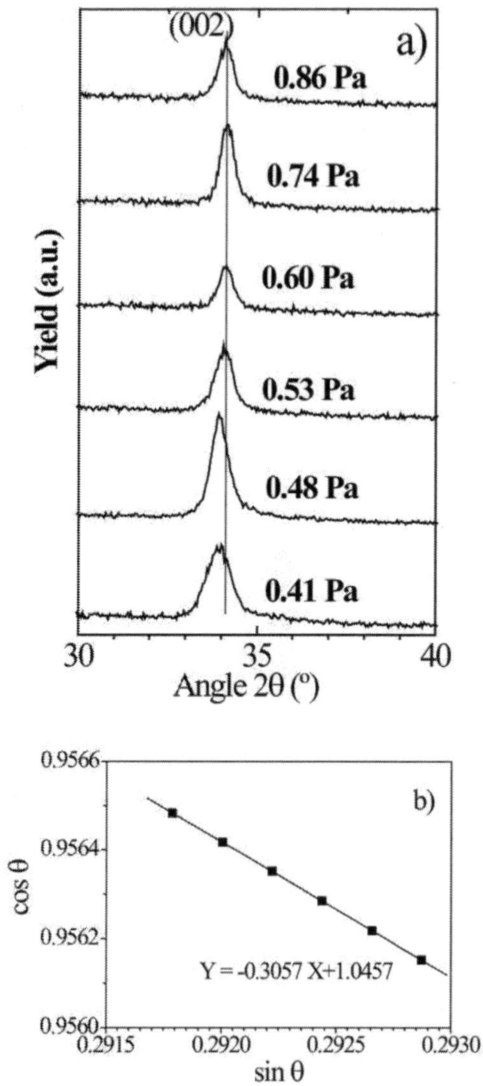

Figure 1 XRD analysis for GZO thin films prepared under different $\boldsymbol{P}_w$s.

3.2 Optical properties

The nc nature of the thin films influences both optical and electrical performance. Figure 2 shows optical transmittance as a function of wavelength for thick GZO films (approx. 700 nm) prepared on glass at various P_w, using air as a reference. The near infra-red transmittance is lower for P_w values of 0.41 and 0.53 Pa and increases with higher P_w, which is consistent with the changes observed in the electrical resistivity (discussed in the next section). The optical band gap for GZO films was calculated by plotting $(\alpha h\nu)^2$ as a function of photon energy ($h\nu$) and extrapolating the linear region of $(\alpha h\nu)^2$ to energy axis where $(\alpha h\nu)^2$corresponds to zero. Figure 2b shows the plot of $(\alpha h\nu)^2$ as a function of photon energy ($h\nu$) for GZO films. From these plots, it can be seen that the value of the bandgap of GZO decreased from 3.73 eV (0.41 Pa) to 3.48 eV (0.86 Pa), which can be understood in the context of the Burstein Moss shift [9].

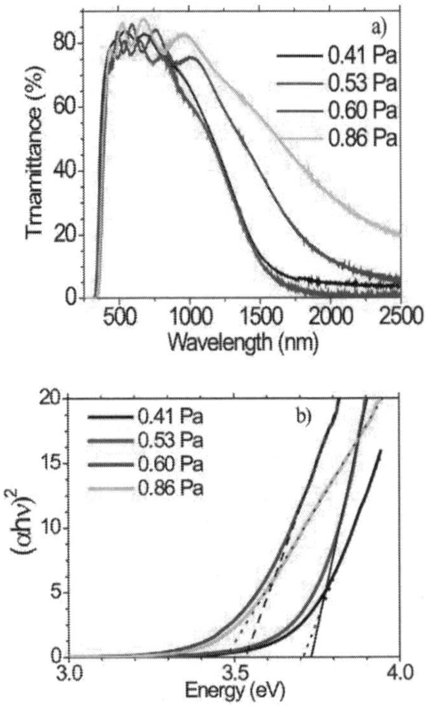

Figure 2 Optical transmittance of GZO/glass at various $\boldsymbol{P}_w$s.

3.3 Electrical properties

The electrical resistivity, charge carrier concentration and Hall mobility as a function of the P_w, for GZO films deposited on glass, are shown in Figure 3. The resistivity of GZO samples decreased initially, and then increased with the P_w. In general, the average resistivity was low (approx. 10^{-4} Ω cm), which can be attributed to high carrier concentration. Considering the similarity in the conduction mechanism of electrons in GZO and ITO, the grain boundary (GB) and ionized impurity scattering processes can be considered the two dominant mechanisms, limiting electron transport in nc-GZO films, as in the case of ITO, where other scattering mechanisms such as lattice vibrations and neutral impurity scattering may typically be neglected [10]. The relative importance of the scattering mechanism is dependent on film quality and carrier concentration. Unlike intrinsic ZnO, where the conduction is generally controlled by GB-scattering, in doped ZnO at high electron density ($>10^{20}$ cm^{-3}), the ionized impurity scattering can be expected to dominate, which explains the low values of electron mobility (<10 cm^2V/s) [11].

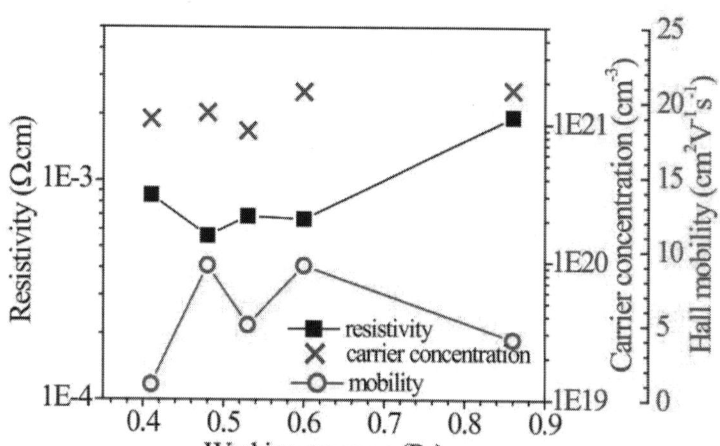

Figure 3 The electrical resistivity, carrier concentration and Hall mobility for GZO/glass as a function of the $\boldsymbol{P}_w$.

3.4 Tensile tests

Tensile tests were performed at a constant strain rate on PEN substrates (82 μm) coated with GZO films (approx. 100 nm) prepared under two different P_ws to increase nominal strains. The PEN substrate is isotropic, and the elastic modulus was 4.23 GPa, as measured through the tensile test on uncoated substrate. The cracking densities as a function of the substrate nominal strain for two different GZO coatings (0.53 and 0.86 Pa) are shown in Figure 4a. The crack densities at saturation of these two PEN/GZO films were 0.316 and 0.515 μm^{-1}, respectively. The coatings have similar properties and thicknesses, with small differences causing variations wholly within acceptable margins of error. Using the weakest link model, the coating's cohesive strength was evaluated from the early stages of the fragmentation process, assuming a Weibull-type, size-dependent probability of failure for the coating fragments of length ℓ under a stress σ [12, 13]:

$$F(\sigma) = 1 - \exp\left[-\frac{\ell}{\ell_0}\left(\frac{\sigma}{\beta}\right)^{\alpha}\right] \qquad (2)$$

Assuming that the residual stresses were negligible, in the initial stage of fragmentation, the average fragment length was related to the stress acting in the coating. The average fragment length (ℓ) is $\ell_0(\sigma/\beta)^{-\alpha}$, where a normalizing factor (ℓ_0) of 1 μm was chosen. In addition, σ is the axial stress acting in the coating, and α and β are the Weibull shape and scale parameters, respectively. These parameters were derived from a plot of ln(ℓ) versus ln(σ), shown in Figure 4b, using the initial part of the crack density evolution of the PEN/GZO coatings, displayed in Figure 4a.

The cohesive strength of the coating at critical length (ℓ_c) can be expressed as

$$\sigma_{max}(\ell_c) = \beta\left(\frac{\ell_c}{\ell_0}\right)^{-1/\alpha}\Gamma(1 + 1/\alpha) \qquad (3)$$

where Γ is the gamma function, $\ell_c = (3/2)\ell_{sat}$ is the critical length and ℓ_{sat} is the experimental mean fragment length at saturation, which is also the inverse of the crack density at saturation [14]. As shown in Figure 4a, the GZO coatings prepared at P_w of 0.53 and 0.86 Pa revealed mean fragment lengths at saturation of 3.11 and 1.94 μm, respectively.

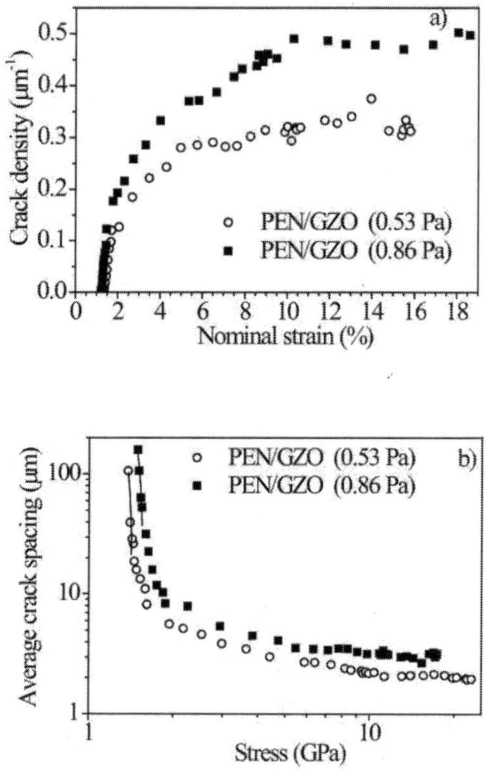

Figure 4 Cracking density as a function of the substrate nominal strain for different GZO coatings deposited on PEN (82 μm) and the crack density evolution of the PEN/GZO coatings.

In order to take into account its influence, the internal stress was evaluated, and the COS and the coating strength obtained with this method were corrected.

$$COS^{cor} = COS + \varepsilon_i \qquad (4)$$

where σ_i is the internal stress and $\varepsilon_i = \sigma_i(1 - \nu_c)/E_c$, the internal strain, with E_c and ν_c being the Young's modulus and Poisson ratio, respectively, of the coating. Young's modulus of GZO was measured by nanoindentation at 113 and 112 GPa from samples prepared at 0.60 and 0.86 Pa, respectively. Young's modulus of the PEN substrate was determined from tensile testing (4.23 GPa). The cohesive strength of the coatings, which was evaluated from the initial part of the crack density evolution, was found to be between 1.3 and 1.4 GPa. The crack onset strains (COS^{cor}) occurs for nominal strains of 1.1 and 1.0%, respectively. The COS and cohesive strength of GZO are relatively similar to those reported in the literature for other polycrystalline conducting films [15].

4. SUMMARY

The material, opto-electrical properties, COS, the coating cohesive strength, as well as the influence of mechanical strain on the electrical properties of nc GZO thin films were investigated. The estimated average crystalline size of nc-GZO films was approx. 8.7 nm, and the bandgap shifted from 3.73 eV (0.41 Pa) to 3.48 eV (0.86 Pa), where the low resistivity (approx. 10^{-4} Ω cm) and the high electron density ($>10^{20}$ cm^{-3}) explain the dominating scattering process as the ionized impurity scattering. The COS is similar for different GZO coatings and occurs for nominal strains approx. 1%. The cohesive strength of coatings, which was evaluated from the initial part of the crack density evolution, was found to be between 1.3 and 1.4 GPa, while the Young's modulus was evaluated by nanoindentation.

ACKNOWLEDGEMENTS

The authors acknowledge the receipt of funding from the Portuguese Foundation for Science and Technology (FCT) Grant PTDC/CTM/ 69316/2006, INL project 156: SIMBIO, NANO/NMed-SD/0156/2007 and the CIENCIA 2007 programme.

COMPETING INTERESTS

The authors declare that they have no competing interests.

AUTHORS' CONTRIBUTIONS

LR and SLM proposed the research work, and with APS coordinated the collaborations and carried out the analysis and interpretation of the experimental results. VG and LRP participated in sample processing, electromechanical experimental measurements, and analysis and interpretation of the results. PA, AVK and YAD carried out electrical measurements and SC performed the nanoindentation measurements. All authors read and approved the final manuscript.

REFERENCES

1. Fonrodona M, Escarré J, Villar F, Soler D, Asensi JM, Bertomeu J, Andreu J: PEN as substrate for new solar cell technologies. *Sol Energy Mater Sol Cells* 2005, 89: 37.

2. Kyaw AKK, Sun XW, Zhao JL, Wang JX, Zhao DW, Wei XF, Liu XW, Demir HV, Wu T: Top-illuminated dye-sensitized solar cells with a room-temperature-processed ZnO photoanode on metal substrates and a Pt-coated Ga-doped ZnO counter electrode. *J Appl Phys D Appl Phys*2011, 44: 045102.

3. Taylor MP, Readey DW, van Hest MFAM, Teplin CW, Alleman JL, Dabney MS, Gedvilas LM, Keyes BM, To B, Perkins JD, Ginley DS: The Remarkable Thermal Stability of Amorphous In-Zn-O Transparent Conductors. *Adv Funct Mater* 2008, 18: 3169.

4. Hamberg I, Granqvist CG: Evaporated Sn-doped In2O3 films: Basic optical properties and applications to energy-efficient windows. *J Appl Phys* 1986, 60: R123.

5. Fortunato E, Gonçalves A, Assunção V, Marques A, Águas H, Pereira L, Ferreira I, Martins R: Growth of ZnO:Ga thin films at room temperature on polymeric substrates: thickness dependence. *Thin Solid Films* 2003, 442: 121.

6. Lewis BG, Paine DC: Applications and Processing of Transparent Conducting Oxides. *MRS Bull* 2000, 25: 22.

7. Gu F, Wang SF, Lu MK, Zhou GJ, Xu D, Yuan DR: Structure Evaluation and Highly Enhanced Luminescence of Dy3+-Doped ZnO Nanocrystals by Li+ Doping via Combustion Method. *Langmuir* 2004, 20: 3528.

8. Cullity BD, Stock SR: *Elements of X-Ray Diffraction*. 3rd edition. NJ: Prentice-Hall Inc; 2001:167–171. ISBN 0–201–61091–4 ISBN 0-201-61091-4

9. Park JB, Park SH, Song PK: Electrical and structural properties of In-doped ZnO films deposited by RF superimposed DC magnetron sputtering system. *J Phys Chem Solids* 2010, 71: 669.

10. Robbins JJ, Harvey J, Leaf J, Fry C, Wolden CA: Transport phenomena in high performance nanocrystalline ZnO:Ga films deposited by plasma-enhanced chemical vapor deposition. *Thin Solid Films* 2005, 473: 35.

11. Minami T: Transport phenomena in high performance nanocrystalline ZnO:Ga films deposited by plasma-enhanced chemical vapor deposition. *MRS Bull* 2000, 25: 38.

12. Weibull W: A statistical distribution function of wide applicability. *J Appl Mech* 1951, 18: 293.

13. Leterrier Y, Boogh L, Andersons J, Månson J-AE: Adhesion of silicon oxide layers on poly(ethylene terephthalate). I: Effect of substrate properties on coating's fragmentation process. *J Polym Sci B Polym Phys* 1997, 35: 1449.

14. Leterrier Y: Durability of nanosized oxygen-barrier coatings on polymers. *Prog Mater Sci* 2003, 48: 1.

15. Leterrier Y, Médico L, Demarco F, Månson J-AE, Betz U, Escola MF, Olsson MK, Atamny F: Mechanical integrity of transparent conductive oxide films for flexible polymer-based displays. *Thin Solid Films* 2004, 460: 156.

10

A STUDY ON THE CHARACTERISTICS OF PLASMA POLYMER THIN FILM WITH CONTROLLED NITROGEN FLOW RATE

Sang-Jin Cho and Jin-Hyo Boo

Department of Chemistry and Institute of Basic Science, Sungkyunkwan University, Suwon 440-746, South Korea

ABSTRACT

Nitrogen-doped thiophene plasma polymer [N-ThioPP] thin films were deposited by radio frequency (13.56 MHz) plasma-enhanced chemical vapor deposition method. Thiophene was used as organic precursor (carbon source) with hydrogen gas as the precursor bubbler gas. Additionally, nitrogen gas [N_2] was used as nitrogen dopant. Furthermore, additional argon was used as a carrier gas. The as-grown polymerized thin films were analyzed using ellipsometry, Fourier-transform infrared [FT-IR] spectroscopy, Raman spectroscopy, and water contact angle measurement. The ellipsometry results showed the refractive index change of the N-ThioPP film. The FT-IR spectra showed that the N-ThioPP films were completely fragmented and polymerized from thiophene.

KEYWORDS

PECVD, plasma polymer, N-ThioPP, optical physical and chemical properties

1. INTRODUCTION

The existing semiconductor technology lets a silicone material make integrations by a top-down form developed by nano- or molecule technology merged with nanotechnology, biotechniques, and information technology and by a bottom-up method to constitute a device and a circuit with self-alignment of atoms and molecules. Those are common opinions of a majority of experts. In spite of the basic consensus by such experts, the progress in the nano and molecule device research field is very slow, and it is much worse now. There are many causes as possible reasons, but it is recognized that the following are still in question: 'the choice of the stable molecule and design technology,' 'self-alignment technology of atoms and molecules,' and 'technology to form a molecule and contact between the metal electrode for stability' [1, 2]. The realization of nanoscale electronics expects the development of bottom-up strategies such as chemical synthesis, self alignment of atoms and molecules, and self-assembled supramolecule. In fabricating the bio-application material, the diamond-like carbon [DLC] films have been a good candidate for some applications such as blood-contacting devices [3] and cell-contacting materials [4] due to their excellent mechanical properties [5–7].

In this work, nitrogen-doped plasma polymer was deposited by nitrogen injection during the plasma-enhanced chemical vapor deposition [PECVD] process without ammonia gas. Also, N-ThioPP thin films were investigated on the surface properties such as surface energy and structural effects.

2. EXPERIMENTAL DETAIL

The experiment was carried out in a homemade stainless-steel PECVD system as shown in Figure 1. Silicon(100) wafers were wet-cleaned by sonication with acetone, ethyl alcohol, distilled water, and isopropyl alcohol and dried by N_2 gas blowing. Also, substrates were dry-cleaned by *in situ* Ar plasma bombardment with 100 W for 15 min. The plasma polymer thin films were deposited by PECVD method. Thiophene was utilized as organic precursor. Thiophene was preheated up to 60°C and bubbled by 50 sccm of hydrogen gas. Additionally, 50 sccm of argon gas was used as a carrier gas. The deposition time was 45 min to make the same thickness at 1 μm. The deposition pressure and temperature were 4.0×10^{-1} Torr and 25°C, respectively. The typical conditions of the PECVD process applied in this study for film deposition are 0, 20, 30, 40, and 50 sccm of N_2 gas flow.

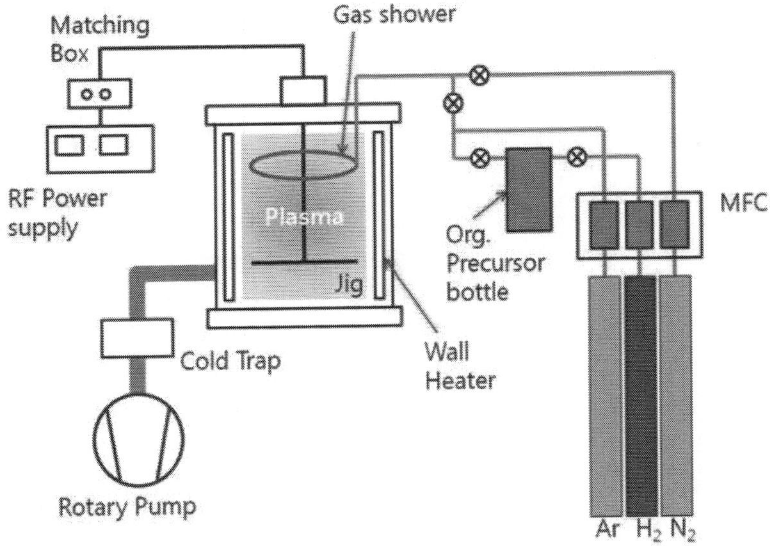

Figure 1 A schematic diagram of the home-made PECVD system.

The chemical bonding type of plasma polymer thin films was investigated by FT-IR spectroscopy (Vertex 70, Bruker Optik Gmbh, Ettlingen, Germany). Moreover, Raman shift of each thin film was investigated by FT-Raman spectroscopy (Vertex 70 with RAM-II, Bruker Optik Gmbh, Ettlingen, Germany). Surface wettability was measured according to water contact angle measurements (Attension, KSV Instruments, Ltd., Helsinki, Finland). The *ex-situ* ellipsometry data of all investigated films were produced by an ellipsometer (GC5A automatic ellipsometry, Gaertner Scientific Corporation, Skokie, IL, USA) at 632 nm to investigate the relationship of film density with the doping amount of nitrogen. Transmittance and bandgap energy of the N-ThioPP thin film were investigated by a UV-Vis spectrophotometer (Optizen 2120UV Plus, Mecasys Co., Ltd., Yuseong-gu, Daejeon, South Korea).

3. RESULTS AND DISCUSSION

The bonding state of the plasma-polymerized thin films was analyzed by FT-IR absorption over a range of 4,000 to 600 cm^{-1}, as shown in Figure 2. Bands from 700 to 1,050; 1,180; 1,440; 1,601; 1,667; 1,700; 2,025; 2,200; 2,800 to 3,000; 3,200 to 3,600; 3,550; and 3,700 cm^{-1} corresponded to the alkenes (CH_x), C = S, CH_x bending vibrations, C = C; C = O, amide, isonitrile (aromatic), nitrile (aromatic), CH_x stretching vibrations, OH, NH_x, and OH bands, respectively [8]. OH absorption bands come from the air during *ex-situ* FT-IR measurement. NH_x and nitrile absorption band were observed with nitrogen-doped samples. Increasing the N_2 flow rate during the PECVD led to an increase in the nitrogen doping amounts and nitrogen species binding in the ThioPP thin film. Also, the shape of the fingerprint region is different between the ThioPP and N-ThioPP thin films. Notably, the amide shoulder peak (in the vicinity of 1,700 cm^{-1}) is observed only in nitrogen-doped ThioPP samples. From those results, nitrogen was bonded with the ThioPP film by nitrogen injection during the PECVD process. Additionally, Raman spectra also show the same evidence as the IR result. The Raman shift of each sample was shown in Figure 3. The assignments of the bands are observed in the vicinity of 3,200 cm^{-1} and 770 cm^{-1} in the NH_x vibrations and CH out-of-plane bend, respectively [9–11]. NH_xpeaks were increased with increasing N_2 flow rate. Also, there are no peaks in the vicinity of 1,300 cm^{-1} and 1,600 cm^{-1}. It means that the N-ThioPP thin film is not a DLC film in this experimental condition [12].

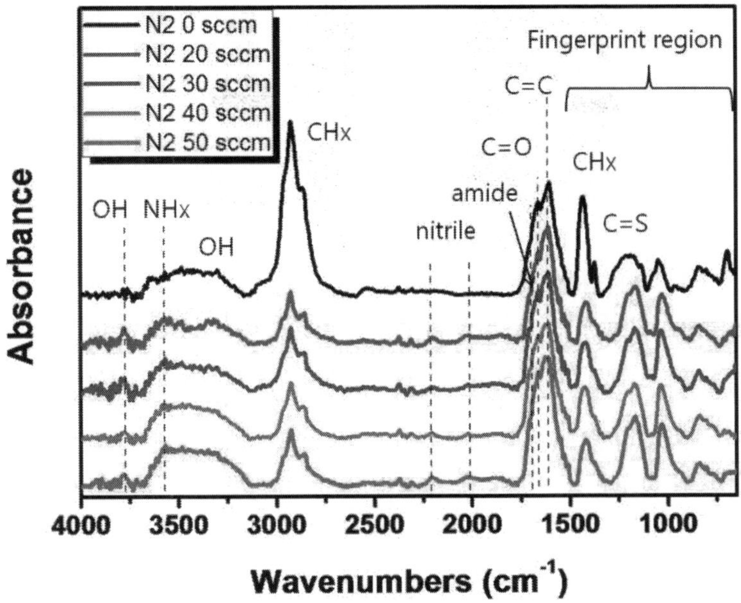

Figure 2 FT-IR spectra of ThioPP and N-ThioPP with N_2 flow rate.

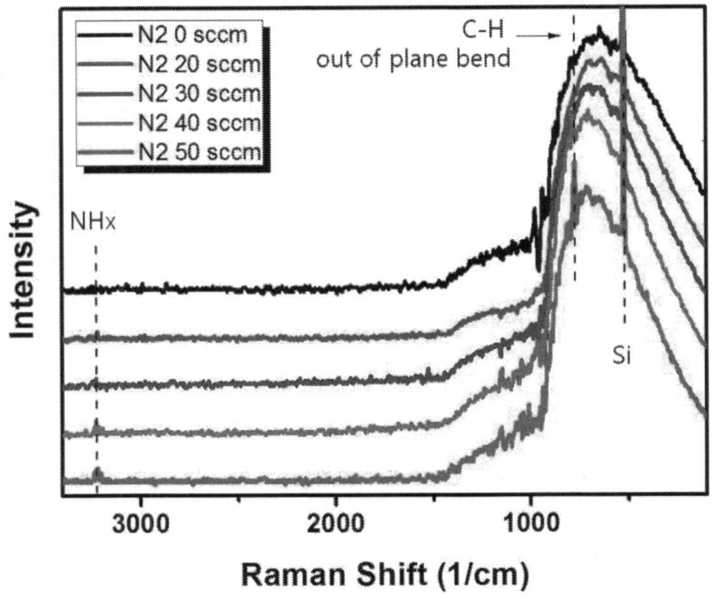

Figure 3 Raman spectra of ThioPP and N-ThioPP with N_2 flow rate.

The change of water contact angles with increasing N_2 flow rate was shown in Figure 4. The water contact angles were decreased by increasing the N_2 flow rate. Surface energy of the ThioPP thin film was changed into a more hydrophilic surface by the high flow rate of nitrogen. When 50 sccm of N_2 gas was inserted during the PECVD process, the water contact angle was 74° which was the lowest value. From the FT-IR spectroscopy, FT-Raman spectroscopy, and contact angle measurements, the chemistry and surface energy of the ThioPP thin film were changed by nitrogen amounts.

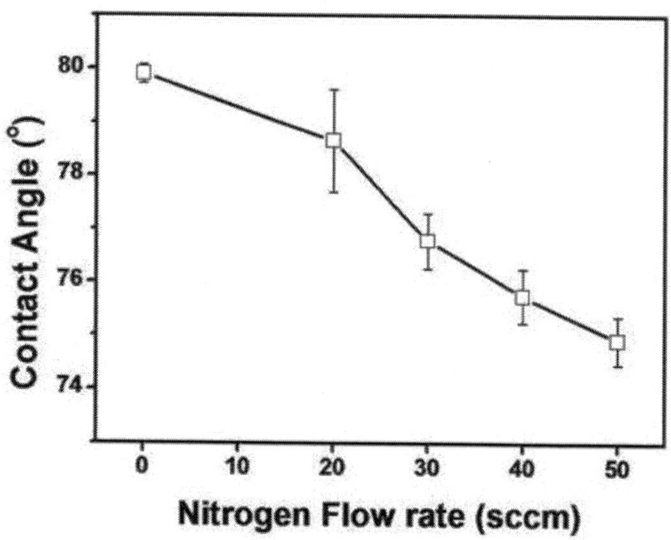

Figure 4 Water contact angle of ThioPP and N-ThioPP with N_2 flow rate.

Figure 5 shows the refractive indices of the plasma polymer thin film with RF power. Refractive index was decreased by increasing the N_2 flow rate. It means that the density of the plasma polymer thin film was decreased by increasing the N_2 flow rate due to the disturbance that led to form a high density cross-link between the thiophene molecules in the plasma polymer. Thus, the refractive index of N-ThioPP thin film was decreased by increasing the N_2 flow rate [13, 14].

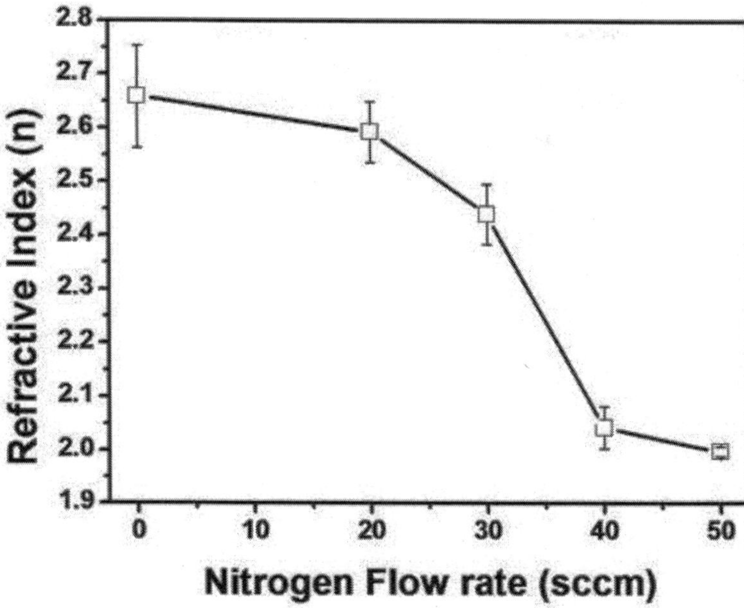

Figure 5 Refractive indices of ThioPP and N-ThioPP with N_2 flow rate.

We measured that the average transmittance in the infrared range was over 80% for all ThioPP films (in Figure 6a). The optical band gap, E_g, of the thin film could be obtained by plotting α^2 vs. $h\nu$ (α is the absorption coefficient and $h\nu$ is the photon energy) and extrapolating the straight-line portion of this plot to the photon energy axis (Figure 6b) [15–18]. As nitrogen contents increased, the absorption edge shifted to a longer wavelength region (Figure 6a). Figure 6b showed the variation of optical band gap as a function of dopant contents, respectively. Optical band gaps were widened with increasing nitrogen flow rate (in Figure6b). The reasons were that the densities of electrons were decreased while nitrogen ions were substituted into carbon sites in the films.

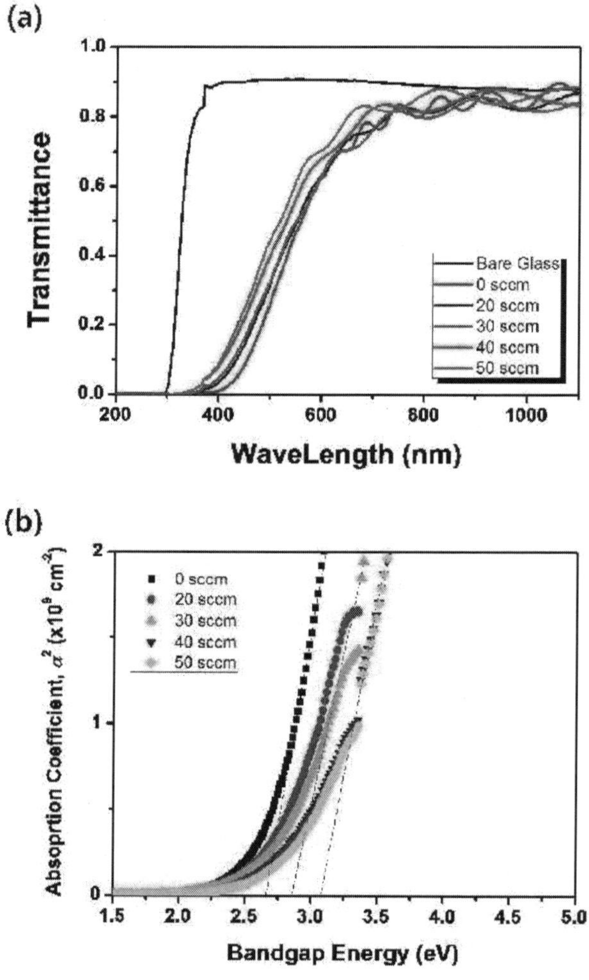

Figure 6 UV-Vis spectra (a) and bandgap energy (b) of ThioPP and N-ThioPP with N_2 flow rate.

4. CONCLUSIONS

N-ThioPP thin films were deposited on Si(100) by the PECVD method. IR spectra (nitrile band, 2,025 and 2,200 cm^{-1}; and the difference of the fingerprint region between ThioPP and N-ThioPP) show that the N-ThioPP thin film was fabricated by nitrogen gas injection during the

PECVD process. Moreover, the increasing NH_x species was definitely shown in the Raman spectra. NH_xspecies was increased by increasing the N_2 flow rate. Also, decreasing the contact angle shows the increasing surface energy of the N-ThioPP thin film with increasing N_2 flow rate. Additionally, decreasing the contact angle indicates the indirect cause of the increasing nitrogen amounts in the N-ThioPP thin film. Nitrogen atoms bonded with thiophene molecules during the PECVD process. Also, nitrogen disturbs the strong bond between thiophene molecules. Thus, the refractive index of the N-ThioPP thin film was decreased by increasing the nitrogen amount. It indicates that the hardness of the N-ThioPP thin film was controlled by nitrogen amounts in the thin film. UV-Vis spectra of all samples show 80% of transmittance in the infrared region. However, transmittance in the visible region was dramatically changed by increasing the nitrogen amounts. Thus, the energy bandgap of N-ThioPP was increased by increasing the nitrogen amounts.

From those results, nitrogen-doped plasma polymer thin films could be fabricated easily by nitrogen injection during the PECVD process without ammonia as toxic gas. Also, we can control the optical, physical, and chemical properties of the N-ThioPP thin film by controlling of nitrogen flow rate.

ACKNOWLEDGEMENTS

This work was supported by NRF-20110027123 (Basic Science Research Program) and NRF-20110000849 (Plasma Bioscience Research Center, SRC Program).

COMPETING INTERESTS

The authors declare that they have no competing interests.

AUTHORS' CONTRIBUTIONS

SJC carried out the thin film deposition and analytical studies, participated in the sequence alignment, and drafted the manuscript. JHB participated in its coordination. All authors read and approved the final manuscript.

REFERENCES

1. Kim HJ, Roh Y, Hong B: Controlled gold nanoparticle assembly on DNA molecule as template for nanowire formation. *J Vac Sci Technol A*2006, 24: 1327–1331.

2. Braun E, Eichen Y, Sivan U, Ben-Uoseph G: DNA-templated assembly and electrode attachment of a conducting silver wire. *Nature* 1998, 391: 775–778.

3. Jones MI, McColl IR, Grant DM, Parker KG, Parker TL: Protein adsorption and platelet attachment and activation, on TiN, TiC, and DLC coatings on titanium for cardiovascular applications. *J Biomed Mater Res* 2000, 52: 413–421.

4. Linder S, Pinkowski W, Aepelbacher M: Adhesion, cytoskeletal architecture and activation status of primary human macrophages on a diamond-like carbon coated surface. *Biomaterials* 2002, 23: 767–773.

5. Eun KY, Lee K-R, Yoon E-S, Kong HS: Effect of polymeric debris on the tribological behavior of diamond-like carbon films. *Surf Coat Technol* 1996, 86–87: 569–574.

6. Kim H-G, Ahn S-H, Kim J-G, Park SJ, Lee K-R: Electrochemical behavior of diamond-like carbon films for biomedical applications. *Thin Solid Film* 2005, 475: 291–297.

7. Grill A: Diamond-like carbon coatings as biocompatible materials--an overview. *Diam Relat Mater* 2003, 12: 166–170.

8. Silverstein M, Webster FX: *Spectrometric Identification of Organic Compounds*. New York: John Wiley & Sons, Inc; 1998.

9. Went GT, Bell AT: Laser Raman spectroscopy of $NH_3$and ND_3adsorbed on TiO (anatase). *Catal Lett* 1991, 11: 111–118.

10. Pittman RM, Bell AT: Raman investigations of NH_3adsorption on TiO_2, Nb_2O_5, and Nb_2O_5/TiO_2. *Catal Lett* 1994, 24: 1–13.

11. Weiss A, Haran G: Time-dependent single-molecule Raman scattering as a probe of surface dynamics. *J Phys Chem B* 2001, 105: 12348–12354.

12. Prawer S, Nugent KW, Lifshitz Y, Lempert GD, Grossman E, Kulik J, Avigal I, Kalish R: Systematic variation of the Raman spectra of DLC films as a function of sp2:sp3 composition. *Diam Relat Mater* 1996, 5: 433–438.

13. Chan K, Gleason KK: Initiated chemical vapor deposition of linear and cross-linked poly(2-hydroxyethyl methacrylate) for use as thin-film hydrogels. *Langmuir* 2005, 21: 8930–8939.

14. Huang YY, Sarkar A, Schultz PC: Relationship between composition, density and refractive index for germania silica glasses. *J Non-Cryst Solids* 1978, 27: 29.

15. Jeong SH, Park BN, Lee S-B, Boo J-H: Metal-doped ZnO thin films: synthesis and characterizations. *Surf & Coat Tech* 2007, 201: 5318–5322.

16. Park KC, Ma DY, Kim KH: The physical properties of Al-doped zinc oxide films prepared by RF magnetron sputtering. *Thin Solid Films*1997, 305: 201–209.

17. Sernelius BE: Self-energy shift in heavily doped, polar semiconductors. *Phys Rev B* 1987, 36: 4878–4887.

18. Hamberg I, Granqvist CG, Berggen KF, Sernelius BE, Engström L: Band-gap widening in heavily Sn-doped In_2O_3. *Phys Rev B* 1984, 30: 3240–3249.

11

MAGNETOELECTRICITY IN $COFE_2O_4$ NANOCRYSTAL-P(VDF-HFP) THIN FILMS

Xiaohua Liu, Shuangyi Liu, Myung-Geun Han, Lukas Zhao, Haiming Deng, Jackie Li, Yimei Zhu, Lia Krusin-Elbaum and Stephen O'Brien

1 Department of Chemistry, The City College of New York, Marshak-1326, 160 Convent Ave, New York, NY 10031, USA.

2 Energy Institute, The City University of New York, New York, NY 10031, USA.

3 Department of Chemistry, The Graduate Center of CUNY, New York, NY 10016, USA.

4 Department of Mechanical Engineering, The City College of New York, Marshak-1326, 160 Convent Ave, New York, NY 10031, USA.

5 Department of Material Science and Condensed Matter Physics, Brookhaven National Laboratory, Building 480, Upton, NY 11973, USA.

6 Department of Physics, The City College of New York, 160 Convent Ave, New York, NY 10031, USA.

ABSTRACT

Transition metal ferrites such as $CoFe_2O_4$, possessing a large magnetostriction coefficient and high Curie temperature (T_c > 600 K), are excellent candidates for creating magnetic order at the nanoscale and provide a pathway to the fabrication of uniform particle-matrix films with optimized potential for magnetoelectric coupling. Here, a series of 0–3 type nanocomposite thin films composed of ferrimagnetic cobalt ferrite nanocrystals (8 to 18 nm) and a ferroelectric/piezoelectric polymer poly(vinylidene fluoride-co-hexafluoropropene), P(VDF-HFP), were prepared by multiple spin coating and cast coating over a thickness range of 200 nm to 1.6 μm. We describe the synthesis and structural characterization of the nanocrystals and composite films by XRD, TEM, HRTEM, STEM, and SEM, as well as dielectric and magnetic properties, in order to identify evidence of cooperative interactions between the two phases. The $CoFe_2O_4$ polymer nanocomposite thin films exhibit composition-dependent effective permittivity, loss tangent, and specific saturation magnetization (M_s). An enhancement of the effective permittivity and saturation magnetization of the $CoFe_2O_4$-P(VDF-HFP) films was observed and directly compared with $CoFe_2O_4$-polyvinylpyrrolidone, a non-ferroelectric polymer-based nanocomposite prepared by the same method. The comparison provided evidence for the observation of a magnetoelectric effect in the case of $CoFe_2O_4$-P(VDF-HFP), attributed to a magnetostrictive/ piezoelectric interaction. An enhancement of M_s up to +20.7% was observed at room temperature in the case of the 10 wt.% $CoFe_2O_4$-P(VDF-HFP) sample.

KEYWORDS

Thin film, Magnetoelectric nanocomposite, Magnetostrictive, P(VDF-HFP), $CoFe_2O_4$

1. BACKGROUND

Magnetoelectric materials, possessing spontaneous electric and magnetic ordering, show applications in multiple-state memory elements, magnetic field sensors, phase shifters, and microwave frequency transducers. Single-phase multiferroics, such as $BiFeO_3$[1], $YMnO_3$[2], and $CdCr_2S_4$[3], exhibit intrinsic magnetoelectric (ME) effect with inherent cross-coupling between magnetic and electric orders. However, such materials are empirically rare[4] and magnetoelectrically weak due to the contraindication between ferroelectricity and magnetism[5]. In addition, the observed ME effect is far below room temperature[6], which severely limits practical use in device fabrication. An alternative approach is two-phase composite films, in which the ferri/ferromagnetic and ferro/piezoelectric phases are combined. With various connectivity schemes (e.g., 0–3 particulate type, 2–2 laminate type, and 1–3 fiber/rod type), these heterostructures have offered the opportunity to tune ferroelectric and magnetic properties independently, and the ME coefficient is 3 orders of magnitude higher than their single-phase counterparts[7].

The magnetoelectric effect in most multiferroic composites is known as strain-mediated, in which the ME coupling is a concerted result of the piezoelectric effect from the piezoelectric phase and magnetostrictive effect from the magnetic phase. An electric field induces a distortion of the piezoelectric phase, which in turn distorts the magnetostrictive phase, generating a magnetic field and vice versa. Substantial ME coupling requires the ferroelectric phase to be in possession of a high piezoelectric coefficient, while the magnetic phase possess both high magnetostriction and resistivity, with an intimate mechanical contact between the two[8]. Ceramic composites have a combination of ferroelectric and magnetic oxides; polymer composites have the magnetic oxides embedded in ferroelectric polymer matrix. The former is limited by high dielectric loss resulting from the interface; the latter offers mechanical flexibility with facile processing. For instance, with high strength and good stability[9], polyvinylidene difluoride (PVDF) and its copolymers such as poly(vinylidenefluoride-co-trifluoroethylene) (P(VDF-TrFE))[10] and poly(vinylidene fluoride-hexafluoropropylene) (P(VDF-HFP))[11–13] are well known for their ferroelectricity and piezoelectricity, which make them ideal candidates for multiferroic film fabrication and ME effect exploration. Transition metal ferrites such as $CoFe_2O_4$, possessing a large magnetostriction coefficient ($\lambda \approx 10^{-4}$)[14] and high Curie temperature (T_c > 600 K)[15], serve as excellent candidates for the magnetic phase.

Although the mechanism of the magnetoelectric coupling is straightforward, complications arise when quantifying the details of polymer-based nanocomposites. The presence of polymorphism (e.g., α, β, γ, δ phases in PVDF), domain walls, grain boundaries, residual stain/magnetization, surface charge, and voids can significantly hinder the ME effect. Andrew and Clarke[16] found that the inclusion of well-dispersed $Ni_{0.5}Zn_{0.5}Fe_2O_4$ nanoparticles in a PVDF matrix can enhance the ferroelectric phase content. Liu et al.[17]. reported epitaxial $BaTiO_3$-$CoFe_2O_4$ nanocomposite thin films (thickness, 100 nm) with phase transition mediated by tensile strain. Recently, a magnetoelectric coupling coefficient of 12 V/cm · Oe was obtained for P(VDF-HFP)/Metglas laminates[18]. Martins et al.[19] fabricated ferrites/PVDF nanocomposites films with thickness of 40 to 50 μm by solvent casting and melt processing. Guo and co-workers prepared particulate $Ni_{0.5}Zn_{0.5}Fe_2O_4$/P(VDF-TrFE) films (thickness, 50 to 60 μm) by wet chemistry routes, with observable magnetodielectric coefficients represented as a percentage change of dielectric constant with and without an external magnetic field[20].

In this work, we report a novel approach to fabricate 0–3 type particulate nanocomposite thin films composed of dispersed $CoFe_2O_4$ (CFO) nanoparticles embedded in P(VDF-HFP) matrix. Prepared through spin/cast-coating techniques, such films exhibit homogenous thickness ranging from 200 nm to 1.6 μm. With a focus on the potential for magnetoelectric coupling, the morphology, microstructure, dielectric, magnetic, and magnetoelectric properties are investigated systematically.

2. METHODS

The $CoFe_2O_4$ nanocrystals were synthesized by a hydrothermal route[21]. In a typical reaction, 2 mmol $Co(NO_3)_2 \cdot 6H_2O$ (Aldrich, 98+%) and 4 mmol $Fe(NO_3)_3 \cdot 9H_2O$ (Aldrich, 98+%) were first dissolved in deionized water. Ethanolamine was dropwise added in the solution until precipitation completed. The obtained precipitate was collected by centrifugation and washed with deionized water. Ammonium hydroxide was then added to re-dissolve the solids. The reaction mixture was transferred into a stainless steel autoclave, with

80% volume filled with the ammonium hydroxide solution. The autoclave was then heated at 200°C for 10 to 30 h. The resultant $CoFe_2O_4$ nanopowders were washed, collected, and dried in air at 60°C overnight.

The $CoFe_2O_4$/polymer nanostructured films were prepared via multiple spin coating and cast coating followed by thermal treatment. N,N-dimethylformamide was first used to dissolve $CoFe_2O_4$ nanoparticles and P(VDF-HFP) pallets or polyvinylpyrrolidone (PVP) powder separately, with concentration of 20 mg/ml. Then, the two suspensions were mixed under ultrasonification, according to the weight ratio of CFO versus polymer, and spin-coated or cast-coated on Si or glass substrates and dried at 90°C under vacuum. The thickness of the obtained thin films (200 nm to 1.6 μm) was controlled by the times and/or rotation speed (300 to 1000 rpm) of the spin coating. To measure film thickness, scanning electron microscopy (SEM) cross-sectional analysis was applied. The Si substrate was scored and cut/fractured in order to observe film cross sections, which were then easily analyzed by SEM. Correct instrumental calibration and review of the film over several regions confirmed thin film uniformity, expected for spin/cast coating, and thicknesses could be determined to within ±7%. For dielectric measurements, the glass substrates were pre-deposited with rectangular (1 mm × 5 mm) Ag bottom electrodes by a thermal evaporator. Top electrodes were deposited (5 mm × 1 mm) after the films were coated and dried, leaving the composite sandwiched between two electrodes with square crossed area of 1 mm × 1 mm.

The phase purity and crystal structure of the $CoFe_2O_4$ particles was analyzed by X-ray diffraction (XRD) with a PANalytical powder X-ray diffractometer (Almelo, The Netherlands) with Ni-filtered Cu $K\alpha$ radiation (λ = 1.54056 Å). The data were collected in step scanning mode from 20° to 80°. Microstructural characterization of the CFO powders was performed by transmission electron microscopy (TEM) with a JEOL 3000 F (Akishima-shi, Japan) with an accelerating voltage of 300 kV. We used a JEOL ARM 200CF equipped with cold field emission gun and spherical aberration correctors for both scanning transmission electron microscopy (STEM) and high-resolution transmission electron microscopy (HRTEM). Surface morphology, nanoparticle distribution, and film thickness of the CFO/polymer composite were evaluated by a

Zeiss Supra 55VP SEM (Oberkochen, Germany). Dielectric measurements including frequency dependence of ϵ', dielectric constant and tan δ, and dielectric loss were measured by an Agilent 4294A precision impedance analyzer. Magnetic measurements including zero field-cooled and field-cooled (ZFC/FC) low field magnetization versus temperature and room temperature hysteresis loops were carried out using a Quantum Design MPMS XL-5 SQUID magnetometer (San Diego, CA, USA), with applied fields up to 5 T and temperatures from 1.84 to 400 K.

3. RESULTS AND DISCUSSION

Highly crystalline nanocrystals with a relatively narrow size distribution and reduced tendency toward aggregation were prepared for the purpose of generating a homogeneous 0–3 nanocomposite structure. Emphasis was on reducing the amount of surface passivation in the form of ligands, in order to optimize surface contact and therefore interaction with the ferroelectric polymer, following formation of the nanocomposite. The balance is in maintaining a highly disperse solvent suspension of the nanocrystals during combination with the polymer (which is aided by surface ligands) and obtaining a physical interaction between nanoparticle and polymer (hindered by long chain alkyl ligands and other typical reagents). Representative transmission electron micrograph (TEM, Figure 1a) illustrates that the samples consist of discrete, nanosized $CoFe_2O_4$ crystals with diameter of 8 to 18 nm. The particles are mostly spherical in shape and exhibit low size distribution. Following solvent evaporation, loose and localized aggregation occurs, possibly due to weak intermolecular interactions common and/or magnetic attraction amongst the nanoparticles. The chemical composition was obtained using energy-dispersive X-ray spectroscopy (EDX or EDS, Figure 1b): the ratio of the peaks is in good agreement with expected elemental composition. The average size determined by statistical analysis of the TEM images is consistent with that calculated by the Scherrer equation[18] from the XRD patterns (Figure 1c), indicating single crystallinity of the CFO nanoparticles. The position and relative intensity of all reflection peaks match well the cubic inverse spinel $CoFe_2O_4$ structure (PCPDS no. 04-006-4148), without indication of crystalline byproducts. The average diameter of the nanocrystals, calculated by the Scherrer formula[22] was determined to be 9 and 11 nm for 10- and 20-h treatments, respectively, indicating a slight increase in the average diameter with longer thermal treatment. The increase in particle dimension is ascribed to the longer reaction time, which allows and promotes the crystal growth after nucleation in the hydrothermal process. Images of isolated nanocrystals at higher magnification (HRTEM, Figure 1d) further confirm the crystallinity and phase purity of the as-synthesized cobalt ferrites. The well-defined two-dimensional lattice fringes of 10-nm nanocrystal indicate good crystallinity and lack of structural defects. The plane distance is measured as 2.99 Å, in good agreement with the (220) interplane spacing of the reported $CoFe_2O_4$ lattice.

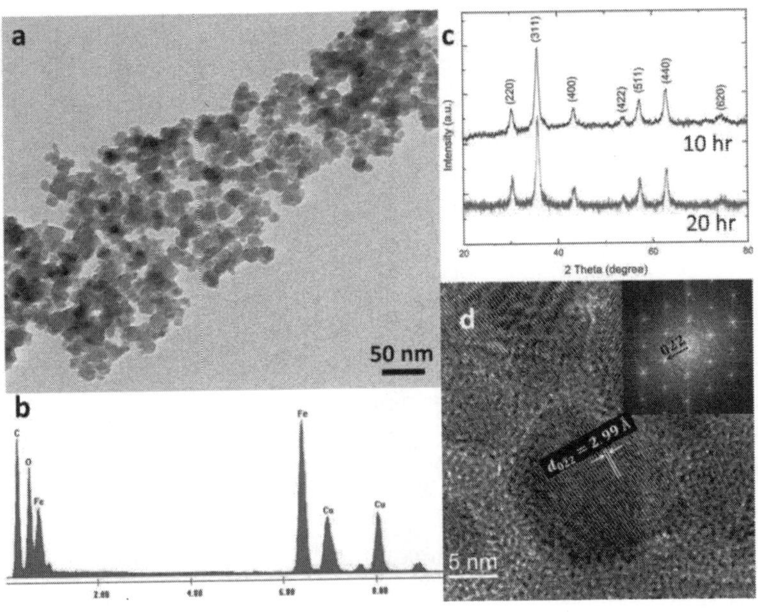

Figure 1 TEM image, EDX spectra, XRD pattern, and HRTEM of $CoFe_2O_4$ nanocrystals. Low magnification TEM image **(a)** of $CoFe_2O_4$nanocrystals synthesized via a solvothermal process and its corresponding EDX spectra **(b)**. **(c)** XRD patterns of the $CoFe_2O_4$nanocrystals reacted for 10 and 20 h. **(d)** High-resolution TEM image. Inset, corresponding its fast Fourier transform indicating the particle is oriented along the zone axis [100].

Considering that the magnetic properties of the nanocrystal were to be compared that of the known bulk behavior of $CoFe_2O_4$, unequivocal identification of the crystal phase, symmetry, and composition of an individual nanocrystal was highly desirable. To further verify the crystal structure, the samples were studied by high angle annular dark field (HAADF) STEM and compared with a calculated model. Figure 2a illustrates the projection of the atomic structure model of $CoFe_2O_4$ along the <110 > zone axis, with oxygen atoms removed. Figure 2b shows the HAADF-STEM image of the as-synthesized nanocrystals, where the bright dots are Co and Fe atoms. The calculated positions of the transition metal atoms are superposed on the HAADF-STEM image, indicating that the elements and positions suggested in the model precisely fit those observed by STEM. As the intensity of the STEM pattern is proportional to Z^2[23], where Z is the atomic number, O atoms are not visible, while Co and Fe atoms are present. Since the atomic numbers of Co ($Z = 27$) and Fe ($Z = 26$) are similar, it would be difficult to distinguish one from the other in the HAADF-STEM image. However, some Co columns exhibit stronger contrast than other Co/Fe columns in Figure 2b. This is because the former Co columns have twice the number of Co atoms as the dimmer ones. In addition, the measured interplane distance of (111) planes (4.80 Å) is consistent with the reported $CoFe_2O_4$ crystal information.

Continuous, uniform, and crack/void-free $CoFe_2O_4$/polymer films with thicknesses in the range 200 nm to 1.6 μm were systematically prepared by multiple spin/cast coating followed by thermal treatment to dry the film. Figure 3 shows SEM images with a CFO weight fraction of 25% where the white dots are the CFO nanoparticles and the dark background is the P(VDF-HFP) copolymer. The top surface view of the microstructure of the nanocomposite film demonstrates that monodisperse, ultrafine cobalt ferrite nanoparticles are well embedded in the polymer matrix, forming typical 0–3, particulate type nanocomposites. Loose agglomeration occurs locally due to the magnetic interaction among the nanopowders. Defects, pores, or phase separation unfavorable for device fabrication was not observed. The cross-sectional image (Figure 3b) confirms the thickness of the free standing film of approximately 1.5 μm. The observation of intimate physical contact between the CFO and P(VDF-HFP) phase components is a good starting point for attempting to generate mechanical, magnetic, or electrical coupling between them.

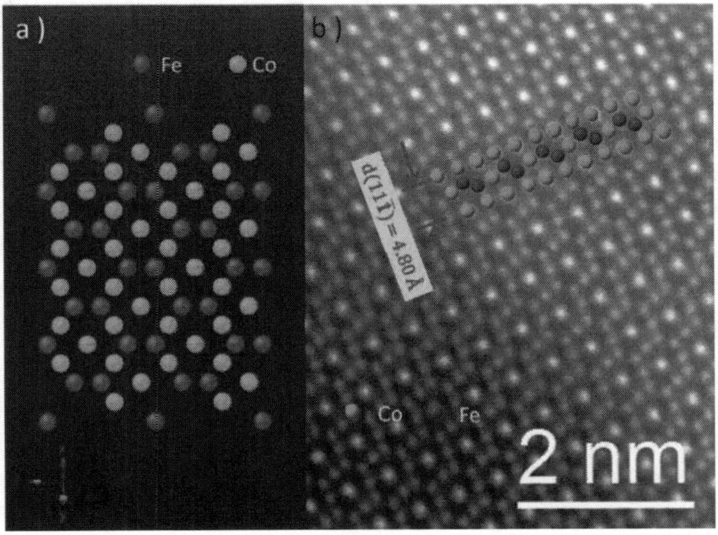

Figure 2 Projection of the inverse spinel structure and the HAADF-STEM image of $CoFe_2O_4$ nanoparticles. **(a)** Projection of the inverse spinel structure of $CoFe_2O_4$ along the <110> zone axis. Red balls represent iron atoms; green balls represent cobalt atoms; oxygen atoms have been removed for clarity. **(b)** Atomic resolution HAADF-STEM image of $CoFe_2O_4$ nanoparticles. Bright balls correspond to cobalt and ferrite atoms. Superimposed points are Co and Fe positions calculated from the crystal model.

The effective permittivity (ϵ_{eff}) and loss tangent (tan δ) of the ferrites/polymer thin films (thickness of approximately 1 μm) were measured over the frequency range from 100 Hz to 1 MHz (Figure 4). Both the effective permittivity and loss tangent of the nanostructured films show a systemic increase as a function of the loading of CFO nanocrystals. The dielectric constant of the pure P(VDF-HFP) film is measured to be 8 at 100 Hz (Figure 4a), consistent with the reported data[24, 25], and increases to 44 in the case of the 30 wt.% CFO samples due to the inclusion of the higher dielectric constant magnetic component ($k(CoFe_2O_4) \approx 400$)[26]. The polarization in ferrites originates from the electronic exchange $Fe^{2+} \Leftrightarrow Fe^{3+}$ and hole transfer between $Co^{2+} \Leftrightarrow Co^{3+}$ in the spinel phase, which cannot follow the alternating external field beyond a certain frequency[27]. When the space charge carriers fail to keep up with the field and lag behind the alternation of its direction, the composites' permittivity and loss tangent decrease monotonically with frequency. Once the frequency is over 10 kHz, the relaxation mechanism associated with the P(VDF-HFP) phase dominates the overall dielectric behavior[20]. The decrease in loss (Figure 4b) with frequency at low frequencies (<1 kHz) is attributed to the ionic DC conduction contribution from the P(VDF-HFP) copolymer phase, which yields interfacial or spatial charge polarization[28]. The increase in loss at high frequencies (>10 kHz) results from the β relaxation associated with the glass transition of the copolymer. Figure 4c further demonstrates the variation of dielectric constant as a function of CFO content in P(VDF-HFP). The nanocomposites show a low composition dependency at higher frequencies, since the dielectric behavior is dominated by the copolymer phase. The PVP films exhibit lower dielectric permittivity (Figure 4d) because the PVP polymer possesses a lower intrinsic dielectric constant of 5.1 (at 100 Hz)[29].

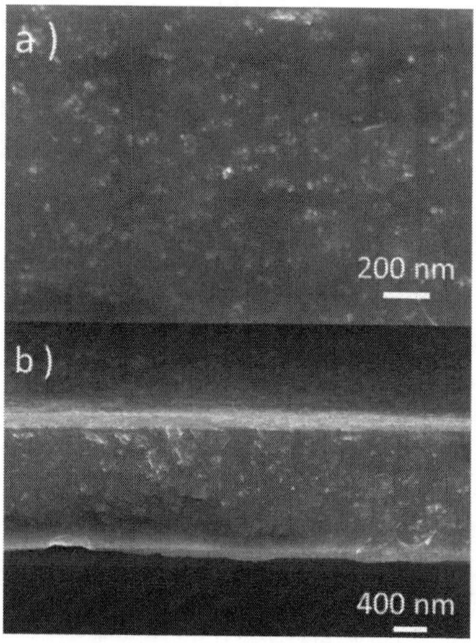

Figure 3 SEM images of $CoFe_2O_4$ / P (VDF-HFP) thin-films deposited on Si substrate. With cobalt ferrite fraction of 25 wt.% and film thickness of 1.5 μm. **(a)** Top surface view; **(b)** cross-sectional view.

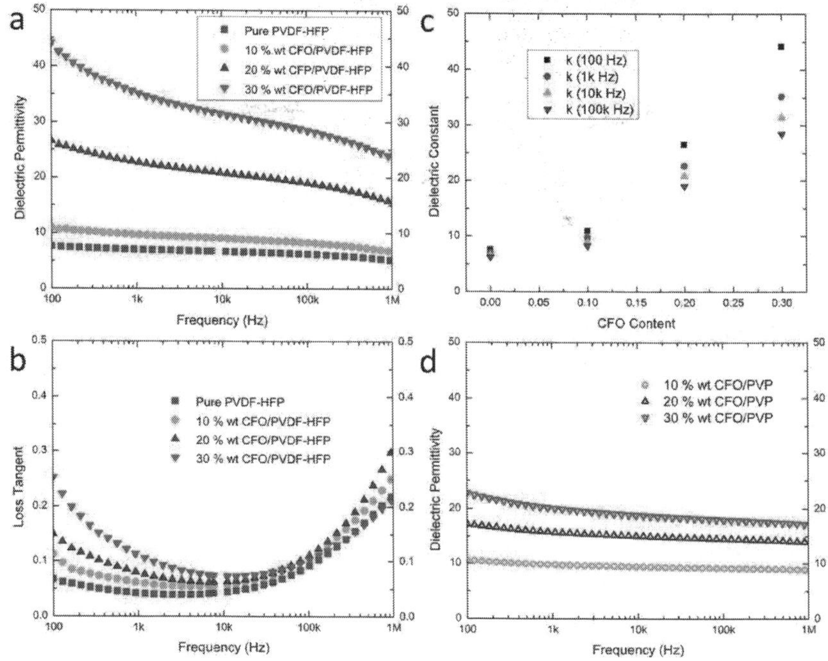

Figure 4 Effective permittivity and loss tangent of the ferrites / polymer thin films. Effective permittivity **(a)** and loss tangent **(b)** of CFO/P (VDF-HFP) nanocomposite thin films with CFO fractions from 0 to 30 wt.%. **(c)** Effective permittivity of the CFO/P(VDF-HFP) as a function of composition at 100 to 1 MHz. **(d)** Effective permittivity of CFO/PVP films.

For 0–3 type nanocomposites with high permittivity nanocrystal fillers discretely distributed in a ferroelectric polymer matrix, the effective permittivity of the films is calculated by the modified Kerner model (or Kerner equation)[30, 31] as shown in Equation 1:

$$\varepsilon_{eff} = \frac{\varepsilon_h f_h + \varepsilon_f f_f AB}{f_h + f_f AB} \tag{1}$$

where

$$A = \frac{3\varepsilon_h}{\varepsilon_f + 2\varepsilon_h} \tag{2}$$

and

$$B = 1 + \frac{3f_f(\varepsilon_f - \varepsilon_h)}{\varepsilon_f + 2\varepsilon_h} \quad (3)$$

The effective permittivity of the films, ϵ_{eff}, is predicted using an average of the host and the filler particle permittivities (ϵ_h and ϵ_f), wherein the contributions are weighted by the fraction of each component (f_f for filler and f_h for host, Equation 1). The measured effective permittivities and those calculated from the modified Kerner model for both PVDF-HFP and PVP films are summarized in Table 1.

Table 1 Comparison of effective permittivity of the CFO/polymer films at 100 kHz from experimental and modified Kerner model

Sample	ε_{eff} (measured)	ε_{eff} (calculated from Kerner equation)	$\Delta\varepsilon_{eff}$
P(VDF-HFP) films			
10 wt.% CFO	9.1	7.3	+1.8
20 wt.% CFO	19.08	13.44	+5.64
30 wt.% CFO	28.56	19.71	+8.85
PVP films			
10 wt.% CFO	9.17	8.82	+0.35
20 wt.% CFO	14.59	13.62	+0.97
30 wt.% CFO	18.05	19.90	−1.85

The effective permittivity of the CFO/P(VDF-HFP) films shows a distinctive and continuous increase relative to the theoretical value estimated by the Kerner model, contrary to the expectations based solely on a composited effective dielectric constant. This can be contrasted with CFO/PVP, which shows significantly less deviation between experiment and theory, and follows expected behavior for a simple combination of two components for ϵ_{eff}. This observation, of deviating behavior in the case of CFO/P(VDF-HFP), is interesting and strongly suggests additional interactions between the polymer and nanoparticle. The phenomenon is ascribed to interfacial interactions between the magnetic filler and the piezoelectric matrix. P(VDF-HFP) undergoes lattice distortion under an applied electric field due to the piezoelectric effect, which introduces local stresses and strain at the ferrite-copolymer interface. Since the thermal shrinkage nature of the P(VDF-HFP) makes complete mechanical coverage of the copolymer over the CFO nanocrystals, and both CFO and P(VDF-HFP) are mechanically hard phases, with Young's modulus of 141.6[32] and 237 GPa[33], respectively, the interfacial stress will be inversely applied to the copolymer phase via strong elastic interactions. As a result of the piezoelectric effect, the spatial charges and electric dipoles within the copolymer matrix are redistributed, manifested as variation of effective permittivities from the Kerner model. With higher ferrite contents, the interfacial elastic effect is stronger and leads to a more pronounced departure from the theoretical value.

Magnetic measurements of the $CoFe_2O_4$ nanocrystals were conducted in both ZFC/FC, and hysteresis modes were analyzed. Figure 5a shows the low field (100 Oe) magnetization dependence with temperature (1.84 to 400 K) in ZFC/FC modes. After a ZFC process, the magnetization of the ferrite nanoparticles increases with rising temperature. Unlike other transition metal ferrite nanoparticles (e.g., Fe_3O_4[34], $NiFe_2O_4$[19], and $MnFe_2O_4$[35]), no maximum magnetization is detected in the ZFC process, indicating that the blocking temperature (T_b) of $CoFe_2O_4$ nanoparticles is above 400 K, which is consistent with reported data of $T_b(CoFe_2O_4)$ = 525 K[19]. Additionally, an irreversible magnetic behavior is indicated by the splitting between the ZFC and FC curves. The irreversibility arises from the competition between the energy required for magnetic moment reorientation against the energy barrier associated with magnetoelectricity and the crystalline anisotropy. The field-dependent magnetization at ambient temperature (Figure 5b) shows a hysteresis with coercivity of 400 Oe, suggesting typical ferrimagnetic behavior. The coercivity represents the strength of the field that is needed to surpass the anisotropy barrier. The saturation magnetization (M_s) and remnant magnetization (M_r) is 66 and 10 emu/g, respectively, comparable with $CoFe_2O_4$ nanocrystals obtained by other approaches with similar sizes[15]. The M_s value of 66 emu/g is equivalent to magnetic moment dipole of 21.6 μ_B per cubic cobalt ferrite unit cell, which is 2.7 μ_B from each Co^{2+} ion. Generally Co^{2+} ions can offer three net spin magnetic moments. The lower value of magnetic moment and subsequent saturation magnetization of these CFO nanoparticles typically originates in the high surface area and concurrent surface disorder. At room temperature, the magnetic anisotropy prevents the magnetization direction of the nanocrystals to completely follow the direction of the external magnetic field.

M(H) hysteresis loops of the $CoFe_2O_4$/P(VDF-HFP) and CFO/PVP nanocomposite thin films were recorded under an applied magnetic field up to 50 kOe. Figure 6a shows hysteresis loops of the 30 wt.% $CoFe_2O_4$/PVDF-HFP thin films at various temperatures, indicating typical ferri/ferromagnetic behavior. At 1.9 K, the 30 wt.% CFO/PVDF-HFP sample is not completely saturated at an applied magnetic field up to 50 kOe, while at higher temperatures (100 to 300 K), it gives a saturation magnetization (M_s) of approximately 20 emu/g. The coercivity of the assembly is 400 Oe at 300 K and reaches 13 kOe at 1.9 K. Figure 6b shows the influence of the nanoparticle loading in the copolymer matrix to the saturation magnetization and remnant magnetization (M_r). The increase in CFO phase content (as volume fraction) gives rise to a systematic increase in the overall $M_s$value; the non-magnetic P(VDF-HFP) polymer does not appear to inhibit the interactions of the magnetic polarization in individual nanocrystals. The composite films show the same coercivity, irrespective of the CFO content.

In order to verify the concerted interaction between the magnetic and ferroelectric phases, hysteresis loops of the CFO/PVP nanocomposites were recorded (Figure 7) and compared with those of the CFO/P(VDF-HFP), presented in Table 2. The saturation magnetization of PVP films are lower compared to PVDF-HFP films with the same composition over the entire magnetic field range. The differences are +1.36 and +2.97 emu/g for 10 and 50 wt.% CFO loading, respectively. The change of the M_s values of the nanocomposite films was normalized for weight fraction and analyzed by the following equation:

$$\Delta M_s\% = \frac{\frac{M_s}{f} - M_{s0}}{M_{s0}} \tag{4}$$

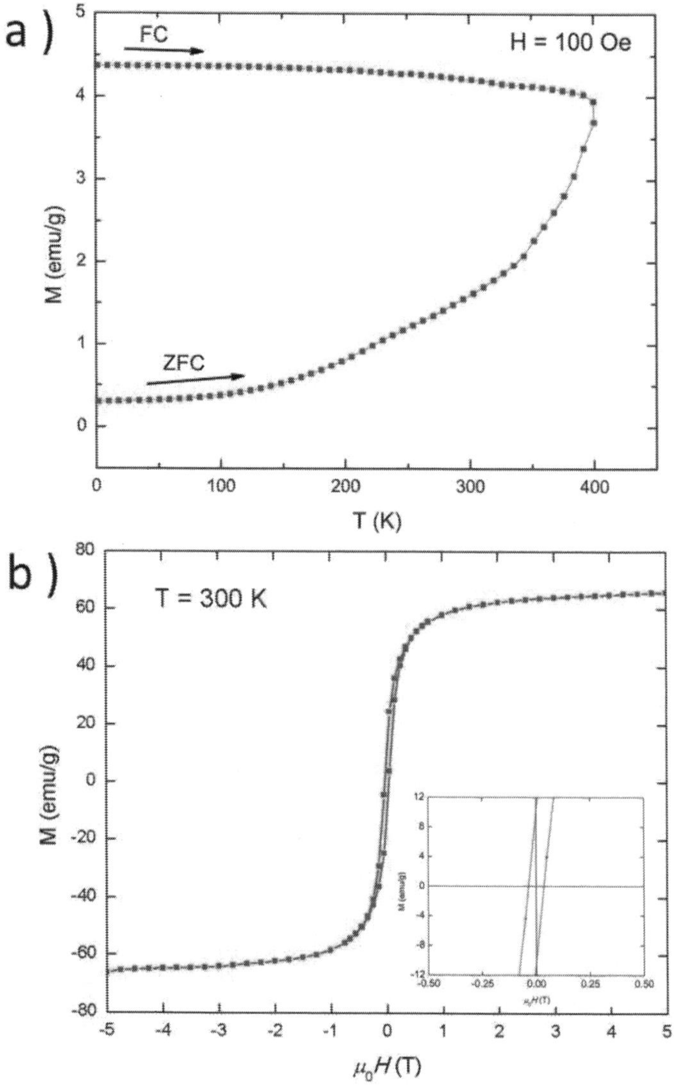

Figure 5 Zero field-cooled and field-cooled (ZFC/FC) and room temperature magnetization curves (a) and hysteresis loop (b).Measured for pure $CoFe_2O_4$ nanoparticles. Inset, central region on an expanded scale.

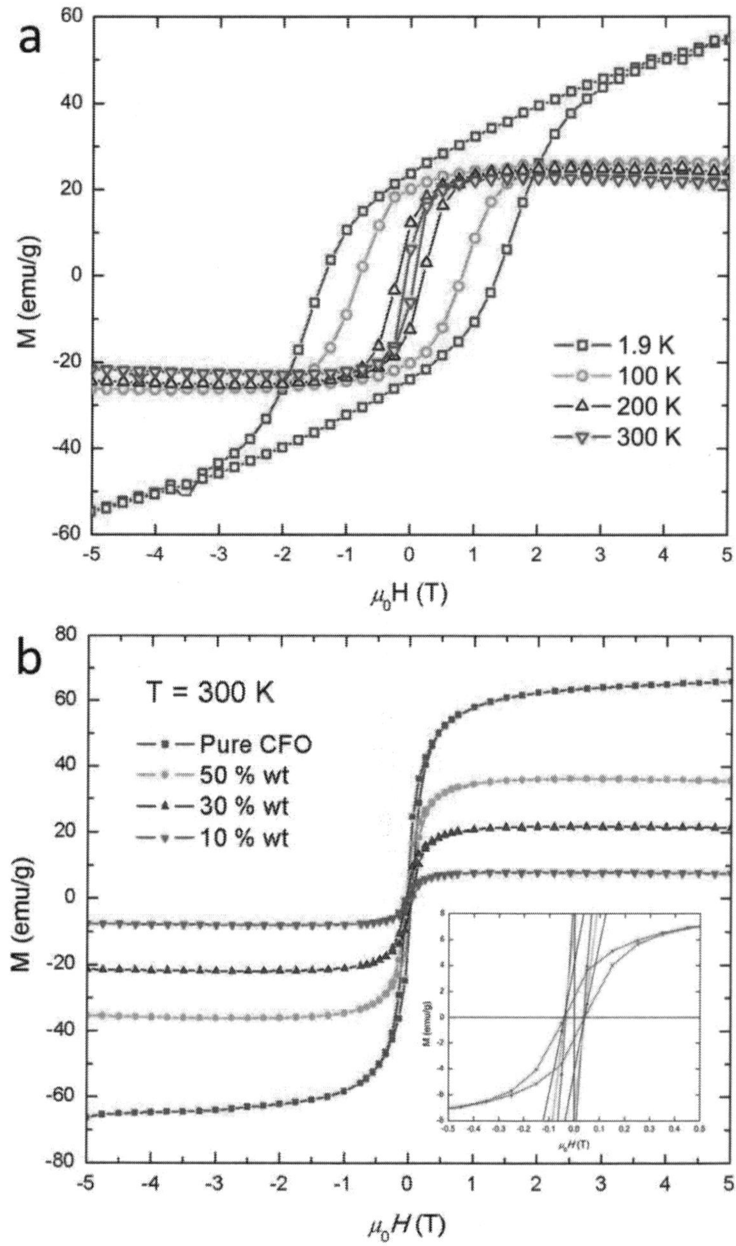

Figure 6 Field-dependent magnetization hysteresis of $CoFe_2O_4$ /P(VDF-HFP) nanocomposites. (a) With 30 wt.% CFO loading at various temperatures and **(b)** at 300 K with various CFO weight fraction. Inset, central region on an expanded scale.

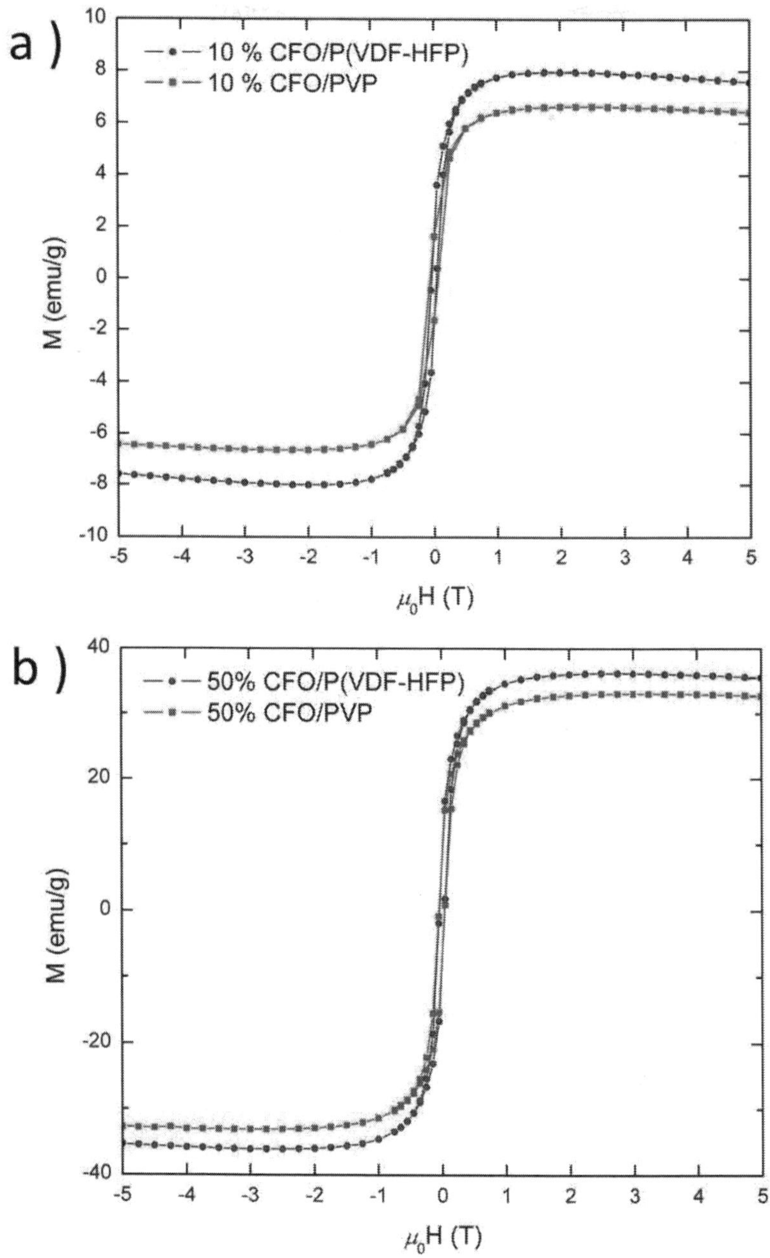

Figure 7 The hysteresis loops of 10 wt.% CFO/P(VDF-HFP) thin-films (a) and 50 wt.% CFO/PVP thin films (b).

Table 2 Saturation magnetization (***M*** $_s$) and normalized percentage change of saturation magnetization (**Δ** ***M*** $_s$ **%**) values for CFO/P(VDF-HFP)and CFO/PVP films with various CFO contents

Sample	M_s (emu/g)	ΔM_s%
P(VDF-HFP) films		
10 wt.% CFO	8.0	+20.7%
30 wt.% CFO	21.8	+9.61%
50 wt.% CFO	36.0	+8.60%
PVP films		
10 wt.% CFO	6.6	+0.09%
30 wt.% CFO	20.2	+0.96%
50 wt.% CFO	33.0	−0.36%

where M_s is the saturation magnetization of a film with certain CFO weight fraction, f is the corresponding weight percentage, $M_{s0}$is the saturation magnetization of pure CFO, and ΔM_s% is the normalized percentage change of the M_s value of each polymer-based film relative to the comparative weighted, pure cobalt ferrite films. The ΔM_s% values for both P(VDF-HFP) and PVP films are summarized in Table 2. The ΔM_s% for the CFO/PVP films is close to zero for all three samples, indicating that the net magnetic moments of the thin films is equivalent to the sum of the contributions from each individual CFO grain inside the PVP matrix (volume fraction contribution only). In contrast, all CFO/P(VDF-HFP) films exhibit positive values of ΔM_s%, with a gradual increase as the copolymer fraction increases. This observation was carefully calibrated against possible errors in the calculation of the volume fraction, most notably through measurement of the film thickness by SEM (see 'Methods' section). This observation strongly suggests a contribution to the ΔM_s due to the presence of P(VDF-HFP) in the form of an enhancement of the saturation magnetization of the composite, with the enhancement stronger for samples with a lower CFO:P(VDF-HFP) ratio.

$CoFe_2O_4$ nanocrystal powders show less than 1% variation in hysteresis loops, whereas CFO/P(VDF-HFP) films show enhancements up to 20.7% in ΔM_S. The enhancement of the M_s value from the P(VDF-HFP) phase, we believe, is a concerted effect and is evident of a ME effect, specifically, through inverse magnetorestrictive coupling. First, the magnetostrictive effect induces a distortion of the crystal lattices of $CoFe_2O_4$ under an applied magnetic field, which in turn leads to local strains or stresses of between the piezoelectric and magnetic phases via intimate mechanical contact. The hypothesis of the influence of intimate mechanical contact between nanocrystals and P(VDF-HFP) is already supported by the observation of permittivity changes unexplained by volume fraction alone, described above. We postulate that the interfacial stress is inversely applied on the CFO phase, which further leads to the change of domain magnetization as a result of an inverse magnetostrictive effect. The effect is quantified by Equation 5:

$$E = \frac{3}{2}\lambda_s \sigma \sin^2\theta \quad (5)$$

where E is the magnetic strain energy density, λ_s is the magnetostrictive expansion at saturation, θ is the angle between the saturation magnetization, and σ is the stress applied on a single magnetic domain[36]. With limited expansion allowed by intimate contact of two hard phases, when compression is applied to CFO phase, the energy is minimized when magnetization is parallel to σ ($\theta = 0$). Consequently, M_s is increased by tension. Moreover, in a sample of pure CFO nanoparticles (M_s = 66 emu/g) each Co^{2+} ion exhibits a magnetic moment of 2.7 μ_B, while in the 10 wt.% CFO/P(VDF-HFP)) films (M_s = 8.0 emu/g), the Co^{2+} ion shows a net magnetic moment of 3 μ_B, which equals the maximum magnetic moment a Co^{2+} ion can offer in the inverse spinel structure. This observation indicates that by intimate mechanical coverage of the CFO particles, P(VDF-HFP) reduces the nanocrystals' degree of surface disorder and surface anisotropy via redistributing charges and dipoles within the copolymer matrix, which allows the magnetization of the cobalt ions to completely follow the external magnetic field. Additionally, as the content of cobalt ferrite nanoparticles increases, the particles' tendency towards agglomeration increases. The interfacial area is reduced due to the formation of small clusters of nanoparticles, and therefore, the interfacial interaction is weakened. This explains why the M_s enhancement is strongest in the 10 wt.% sample (+20.7%), in which the nanoparticles are more completely dispersed, compared to 30 and 50 wt.% samples (+9.6% and +8.6%, respectively).

The magnetoelectric effect associated with the magnetostrictive/ piezoelectric coupling typically can only be observed under high magnetic fields at a very low temperature. In this work, the nanocomposite thin films show substantial magnetoelectric coupling at room temperature. The piezoelectric properties of P(VDF-HFP) and ferrimagnetic properties of $CoFe_2O_4$ nanocrystals are ideal and complimentary in this respect, resulting an observable magnetoelectric coupling.

4. CONCLUSIONS

Crystalline ultrafine CFO with a relatively narrow size distribution from 8 to 18 nm were dispersed in a P(VDF-HFP) copolymer host, forming 0–3 particulate type magnetoelectric nanocomposite thin films. The resulting films exhibit composition-dependent effective permittivity and loss. Following full structural characterization, the magnetic properties of the pure $CoFe_2O_4$nanoparticles were studied and it was confirmed that the saturation magnetization and ZFC/FC curves demonstrate typical ferrimagnetic behavior. By comparing the P(VDF-HFP) and PVP samples, a clear difference in the behavior of the nanocomposite films with respect to effective permittivity and saturation magnetization is observed, highlighting the difference between the use of the ferroelectric polymer and the non-ferroelectric polymer. A magnetoelectric coupling is believed to be observed in the case of CFO/P(VDF-HFP). The origin of the magnetoelectric coupling is attributed to strong elastic interactions between the electric and magnetic phases. The nanocomposite, given its room temperature properties, is an interesting candidate magnetoelectric material with applications in smart devices such as sensors.

ACKNOWLEDGMENTS

This project was supported by the Advanced Research Project Agency for Energy (ARPA-e), ADEPT DE-AR0000114 and the National Science Foundation under award NSF CMMI #1014777. The work was partially funded by the Center for Exploitation of Nanostructures in Sensors and Energy Systems, City College of New York, under NSF Cooperative Agreement award number 0833180. TEM work was supported by the US Department of Energy's Office of Basic Energy Science, Division of Materials Science and Engineering under contract number DE-AC02-98CH10886 and was carried out, in part, at the Center for Functional Nanomaterials, Brookhaven National Laboratory supported by the US Department of Energy, Office of Basic Energy Sciences. Stephen O'Brien acknowledges support from the Columbia-CCNY NSF MIRT, #1122594.

COMPETING INTERESTS

The authors declare that they have no competing interests.

AUTHORS' CONTRIBUTIONS

XL carried out nanoparticle synthesis, thin film fabrication, dielectric properties, and measurements and drafted the manuscript. SL participated in dielectric/magnetic properties characterization and discussion and idea/experiment design. MGH carried out HRTEM and HAADF-STEM analysis, with XL assisting. LZ and HD carried out the magnetic property tests, with XL assisting. JL, YZ, and LKE helped to supervise the experiments and participated in the design of the study and manuscript revision. SO conceived of the study, supervised the project and experiments, and helped to write the manuscript. All authors read and approved the final manuscript.

REFERENCES

1. Wang J, Neaton JB, Zheng H, Nagarajan V, Ogale SB, Liu B, Viehland D, Vaithyanathan V, Schlom DG, Waghmare UV, Spaldin NA, Rabe KM, Wuttig M, Ramesh R: Epitaxial $BiFeO_3$ multiferroic thin film heterostructures. *Science (New York, NY)* 2003, 299: 1719–1722.

2. Lee S, Pirogov A, Han J, Park J-G, Hoshikawa A, Kamiyama T: Direct observation of a coupling between spin, lattice and electric dipole moment in multiferroic $YMnO_3$. *Phys Rev B* 2005, 71: 180413.

3. Sun CP, Huang CL, Lin CC, Her JL, Ho CJ, Lin J-Y, Berger H, Yang HD: Colossal electroresistance and colossal magnetoresistance in spinel multiferroic $CdCr_2S_4$. *Appl Phys Lett* 2010, 96: 122109.

4. Hill NA: Why are there so few magnetic ferroelectrics? *J Phys Chem B* 2000, 104: 6694–6709.

5. Ramesh R, Spaldin NA: Multiferroics : progress and prospects in thin films. *Nat Mater* 2007, 6: 21–29.

6. Ma J, Hu J, Li Z, Nan C-W: Recent progress in multiferroic magnetoelectric composites: from bulk to thin films. *Adv Mater (Deerfield Beach, Fla)* 2011, 23: 1062–1087.

7. Eerenstein W, Mathur ND, Scott JF: Multiferroic and magnetoelectric materials. *Nature* 2006, 442: 759–765.

8. Vaz CAF, Hoffman J, Ahn CH, Ramesh R: Magnetoelectric coupling effects in multiferroic complex oxide composite structures. *Adv Mater (Deerfield Beach, Fla)* 2010, 22: 2900–2918.

9. Lovinger AJ: Ferroelectric polymers. *Science* 1983, 220: 1115–1121.

10. Zhang Q, Bharti V, Zhao X: Giant electrostriction and relaxor ferroelectric behavior in electron-irradiated poly(vinylidene fluoride-trifluoroethylene) copolymer. *Science (New York, NY)* 1998, 280: 2101–2104.

11. Neese B, Wang Y, Chu B, Ren K, Liu S, Zhang QM, Huang C, West J: Piezoelectric responses in poly(vinylidene fluoride/ hexafluoropropylene) copolymers. *Appl Phys Lett* 2007, 90: 242917.

12. Wegener M, Künstler W, Richter K, Gerhard-Multhaupt R: Ferroelectric polarization in stretched piezo- and pyroelectric poly(vinylidene fluoride-hexafluoropropylene) copolymer films. *J Appl Phys* 2002, 92: 7442.

13. He X, Yao K, Gan BK: Phase transition and properties of a ferroelectric poly(vinylidene fluoride-hexafluoropropylene) copolymer. *J Appl Phys* 2005, 97: 084101.

14. Bozorth RM, Elizabeth FT, Albert JW: Anisotropy and magnetostriction of some ferrites. *Phys Rev* 1955, 99: 1788–1798.

15. Zi Z, Sun Y, Zhu X, Yang Z, Dai J, Song W: Synthesis and magnetic properties of $CoFe_2O_4$ ferrite nanoparticles. *J Magn Magn Mater* 2009, 321: 1251–1255.

16. Andrew JS, Clarke DR: Enhanced ferroelectric phase content of polyvinylidene difluoride fibers with the addition of magnetic nanoparticles.*Langmuir: ACS J Surf Colloids* 2008, 24: 8435–8438.

17. Liu B, Sun T, He J, Dravid VP: Sol–gel-derived epitaxial nanocomposite thin films with large sharp magnetoelectric effect. *ACS nano* 2010, 4: 6836–6842.

18. Lu SG, Jin JZ, Zhou X, Fang Z, Wang Q, Zhang QM: Large magnetoelectric coupling coefficient in poly(vinylidene fluoride-hexafluoropropylene)/Metglas laminates. *J Appl Phys* 2011, 110: 104103.

19. Martins P, Costa CM, Botelho G, Lanceros-Mendez S, Barandiaran JM, Gutierrez J: Dielectric and magnetic properties of ferrite/ poly(vinylidene fluoride) nanocomposites. *Mater Chem Phys* 2012, 131: 698–705.

20. Guo Y, Liu Y, Wang J, Withers RL, Chen H, Jin L, Smith P: Giant magnetodielectric effect in 0–3 Ni0.5Zn0.5Fe2O4-Poly(vinylidene-

fluoride) nanocomposite films. *J Phys Chem C* 2010, 114: 13861–13866.

21. Yáñez-Vilar S, Sánchez-Andújar M, Gómez-Aguirre C, Mira J, Señarís-Rodríguez MA, Castro-García S: A simple solvothermal synthesis of MFe_2O_4 (M=Mn, Co and Ni) nanoparticles. *J Solid State Chem* 2009, 182: 2685–2690.

22. Choy Tuck C: *Effective Medium Theory*. Oxford: Clarendon Press; 1999.

23. Zhang X-F, Zhang Z: *Progress in Transmission Electron Microscopy 1: Concepts and Techniques*. New York: Springer; 2001.

24. An N, Liu H, Ding Y, Zhang M, Tang Y: Preparation and electroactive properties of a PVDF/nano-TiO_2 composite film. *Appl Surf Sci* 2011, 257: 3831–3835.

25. Jing X, Shen X, Song H, Song F: Magnetic and dielectric properties of barium ferrite fibers/poly(vinylidene fluoride) composite films. *J Polym Res* 2011, 18: 2017–2021.

26. Bhavikatti AM, Kulkarni S, Lagashetty A: Evaluation of A C conductivity & dielectric behavior of cobalt ferrite. *Int J Eng Sci Technol* 2011, 3: 5985–5991.

27. Gul IH, Maqsood A: Structural, magnetic and electrical properties of cobalt ferrites prepared by the sol–gel route. *J Alloys Compd* 2008, 465: 227–231.

28. Gregorio R: Effect of crystalline phase, orientation and temperature on the dielectric properties of poly(vinylidene fluoride) (PVDF). *J Mater Chem* 1999, 34: 4489–4500.

29. Jung C-H, Cho H, Lee S-Y, Hong Y, Lee C, Hwang D-H: Photo-curable epoxy functionalized cyclotetrasiloxane as a gate dielectric for organic thin film transistors. *Curr Appl Phys* 2010, 10: 1132–1136.

30. Jayasundere N, Smith BV: Dielectric constant for binary piezoelectric 0–3 composites. *J Appl Phys* 1993, 73: 2462.

31. Kim P, Doss NM, Tillotson JP, Hotchkiss PJ, Pan M, Marder SR, Li J, Calame JP, Perry JW: High energy density nanocomposites based on surface-modified $BaTiO_3$ and a ferroelectric polymer. *ACS Nano* 2009, 3: 2581–2592.

32. Zheng H, Wang J, Lofland SE, Ma Z, Mohaddes-Ardabili L, Zhao T, Salamanca-Riba L, Shinde SR, Ogale SB, Bai F, Viehland D, Jia Y, Schlom DG, Wuttig M, Roytburd A, Remesh R: Multiferroic $BaTiO_3$-$CoFe_2O_4$ nanostructures. *Science* 2004, 303: 661–663.

33. Tashiro K, Kobayashi M, Tadokoro H, Fukada E: Calculation of elastic and piezoelectric constants of polymer crystals by a point charge model: application to poly(vinylidene fluoride) form I. *Micromolecules* 1980, 13: 691–698.

34. Adireddy S, Lin C, Palshin V, Dong Y, Cole R, Caruntu G: Size-controlled synthesis of quasi-monodisperse transition-metal ferrite nanocrystals in fatty alcohol solutions. *J Phys Chem C* 2009, 113: 20800–20811.

35. Liu C, Zhang ZJ: Size-dependent superparamagnetic properties of Mn spinel ferrite nanoparticles synthesized from reverse micelles.*Chem of Mater* 2001, 13: 2092–2096.

36. Bozorth RM: *Ferromagnetism*. Princeton: Van Nostrand; 1951:611.

12

POLYMERIZATION OF THIN FILM POLYMERS

Markus Woehrmann[1] and Michael Toepper[2]

[1] Technische Universität Berlin (TUB), Berlin, Germany

[2] Fraunhofer Institute for Reliability and Microintegration (IZM), Berlin, Germany

1. INTRODUCTION

Polymeric coatings such as polyurethanes, acrylic, epoxies and silicones have been used for over 40 years to protect printed wiring boards (PWB) from moisture, handling and environmental influences. Special semiconductor grade polymers have been developed for chip passivation layers. Polyimide became the standard passivation layer for memory chips and other devices needing surface protection for handling and testing procedures. Photosensitive resins have been developed to reduce processing costs. Thin film polymers are used widely in the area of electronic packaging and as an interlayer dielectric (ILD) in nearly every electronic device. Typical layer thickness ranges from 5 µm up to 15 µm. These polymers films should be temperature stable up to range of 150°C (permanent), and for a short time up to 250°C and higher, depending on the soldering process of the components. This is the reason to use thermoset polymers in most cases. The coating process is the spin-coating technology. Spray coating and other techniques are only niche processes. The polymer supplied by the chemical company is a so-called pre-cursor consisting of a partially polymerized polymer in an organic solvent. Nearly all of the thin film polymers need a polymerization step, which is done in most cases by a thermal process after the deposition on the wafer. Polymerization changes the pre-polymer into a long-term stable and much more inherent polymerized dielectric layer. The polymerization process is generally called the *cure* or *cure step* of the process. But the definition "cure" does not mean in every case a fully finished polymerization reaction. Also partly cured polymer films are possible and becoming important if multiple layers have to be deposited. Partly polymerized layers have better adhesion to the following layer compared to fully cured ones. To get something more precise the following definitions have been used: partly cured or soft cured polymers for a 60 – 80 degree of polymerization, and full cured or hard cured for a complete polymerized layer.

There are a huge range of different polymer materials which are used as interlayer dielectric such as polyimides, polybenzoxazole and benzocyclobutene [1]. An example of such a polymer layer on a CMOS-wafer is shown in figure 1:

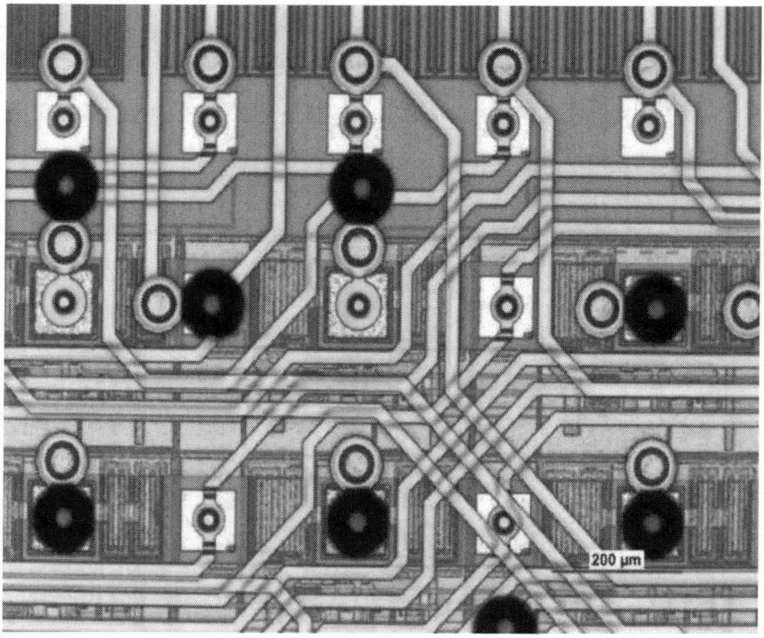

Figure 1. A thin film polymer (BCB from Dow Chemical) as protection layer on a CMOS-wafer

In the last few years, the development of new polymer materials for thin film applications has increased more and more. One main driver for this is the 3-D integration technology which demands optimized material behaviors of the polymers, such as a low coefficient of thermal expansion (CTE) or higher tensile strength. The 3-D technology also demands more and more the low temperature processing. The material combinations get more temperature critical. The reliability will be increased at the same time with low cure temperature, because of a decreased thermal stress in the material stack. In contrast to the demand of low cure temperature, a tough polymer layer is desired with a high decomposition temperature. Analyzing and engineering the polymerization process will help enormous to reach these goals.

This chapter will focus on the benzocyclobutene (BCB) from The Dow Chemical Company being a well-known material for thin-film applications. The full cured BCB has a decomposition temperature well above 350°C. The cure temperature of 250° has been reported in dozens of publications. The analyzing and modeling of the BCB polymerization process will be discussed here in detail. Understanding the behavior of the polymerization process in different states and their kinetic modeling shows the potential for decreasing the cure temperature for a BCB layer which has nearly the same material stability and decomposition temperature as standard processed BCB layers.

2. THE RELEVANCE OF THIN FILM POLYMERS IN MICRO ELECTRONICS

The Dow Chemical Co. (Midland, MI) has developed a variety of low-dielectric (low-k) polymers which are based on different benzocyclobutene monomers (BCB, also known as biscyclo[4.2.0]octa-1,3,5-triene or 1,2-dihydrobenzocyclobutene). Based on these studies, the Dow Chemical Co. commercializes a BCB-Polymer under the registered trademark CYCLOTENE. CYCLOTENE is a family of thermosetting polymers prepared from 1,3 divinyl-1,1,3,3-teramethyldisiloxane-bis-benzocyclobutene (DVS-bis-BCB) monomer. The structure of the DVS-bis-BCB monomer is illustrated in figure 2.

Today CYCLOTENE is just called BCB in most of the literature. BCB has been developed for electronic coating applications. Based on the low-k characteristic of BCB, the material is very attractive as an interlayer dielectric for RF-components [2] [3]. The BCB has been widely used for multilayer re-distribution on wafer level (wafer level packaging-WLP). Since the early 90's, there are hundreds of application notes for BCB (e.g. bumping and redistributing chips, and for planarization and isolation in flat-panel displays, MEMS hermetization, wafer bonding, passivation, gap filling). The wide field of application is based on the material properties such as good planarization, no outgassing, and low copper migration. The Dow Chemical Co. also releases the chemical structure of the monomer and the final polymer, which is not common for the most commercialized thin film polymers. Knowledge of the structure allows for a better understanding of material behavior and material analyzing, which leads to the fact that BCB is used very often for academic research. Most of the newer polymers are years away from such a comprehensive database. BCB is available as photosensitive (CYCLOTENE 4000 series) and non-photosensitive (CYCLOTENE 3000 series) in the form of a spin-on solution. CYCLOTENE can be purchased in different viscosities to obtain a wide film thickness range of <1 μm up to 26 μm (see table 1). CYCLOTENE is shipped in a partly cross-linked state in mesitylene solvent (trimethylbenzene). The non-photosensitive CYCLOTENE can be structured by dry etch processes or laser ablation. On account of this, it is often called dry-etch BCB. The initial extent of polymerization is slightly different when comparing dry etch BCB with photo BCB. The dry etch version is stable at room temperature, in contrast to the photosensitive parts of the 4000 series, which have a shelf life of a week at room temperature, due to the photo-sensitizer. In a frozen state, the durability increases up to a couple of years.

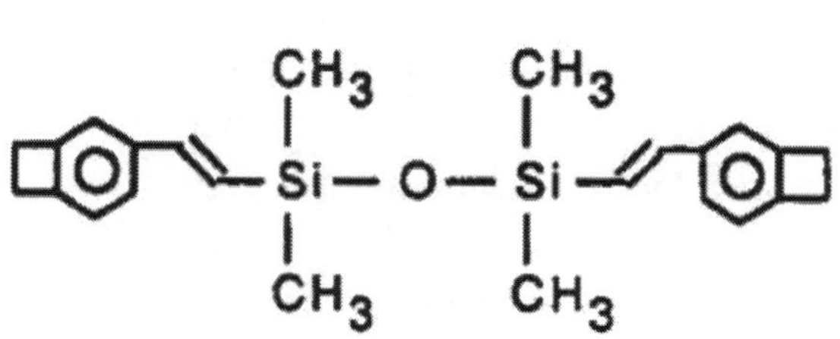

Figure 2. Structure of a DVS-bis-BCB monomer [**4**]

In contrast to polyimides, which are also often used as thin film layers, BCB has a couple of advantages. BCB is highly cross-linked, based on the small monomer, which leads to higher thermal stability in comparison to the cure temperature [4]. It also leads to a complete isotropic behavior of the polymer film. On the other hand, the long-chain characteristic of the polyimide leads to a higher elongation at break, which is an important factor in reliability [1]. The polymerization reaction of the BCB does not emit any volatile by-products, resulting in low cure shrinkage (less than 5 %). The non-polar chemical structure of BCB leads to low dielectric constant (ε_r=2.65) and a high breakdown voltage (530 V/µm) in combination with a low water uptake (less than 0.2 %). The good planarization properties also make BCB attractive for 3-D and adhesive bonding [5] [6].

Table 1. CYCLOTENE with different solid content

Notation	Viscosity (cST@25°C)	Thickness (µm)	Solid content (%)	b-stage (%)
3022-35	14	1.0 – 2.4	35	35
3022-46	52	2.4-5.8	46	35
3022-57	259	5.7-15.6	57	35
3022-63	870	9.5-26.0	63	35
4022-35	192	2.5-5.0	35	47
4024-40	350	3.5-7.5	40	47
4026-46	1100	7.0-14.0	46	47

3. OVERVIEW OF THE POLYMERIZATION PROCESS

3.1. General Polymerization Behavior Of Thermosets

BCB is a thermoset like most other ILD polymers. The polymerization is the process which forms the small molecules or monomers into large polymers structures. The chemical reaction (polymerization) of the thermoset during the processing allows an easy processing of thin film layers. In most cases, a liquid solvent is deposited by spin coating on a substrate, and after the polymerization a tough solid state with a high glass transition temperature is formed. In contrast to thermosets, thermoplastics need a physical condition change for the polymerization. The disadvantage of thermoplastics is the low glass transition temperature ranging from below room temperature up to 180°C, which is a result of the low Van der Waals forces between the polymer chains. For the following processes, such as sputtering, a high thermal stability above 200°C is necessary. A softening of the polymer based on low T_g during sputtering could lead to buckling of the deposited metal layer.

The result of the polymerization of thermosets is a more or less strong 3-D cross-linking of the polymer chains with strong covalent connections in combination with solidification of the polymer layer, which leads to an insolubility of the material. In most cases, thermal energy is used for the polymerization reaction, but there are also other methods like radiation or hardener materials which work like a catalyzer being used for other polymers.

The polymerization is not a monotonic reaction, because it progresses through different stages which change the characteristics of the reaction significantly. The reaction starting point is an ideal solvent in which there are only monomer structures in the solvent. The material is in a "liquid phase" (also called an a-stage). With the supply of enough energy in the system, the cross-linking of the monomers starts and polymer chains build up. The degree of cure describes the rate between reacted parts and unreacted parts (reactants and products). In literature there is a different notation which means more or less the same thing, like extent of cure, extent of reaction, extent of conversion and fraction transformation. The reaction rate is nearly constant in a viscous liquid stage up to the gelation point, which described the abrupt and irreversible transformation of the material into an elastic gel or rubber, and the material loses its ability to flow. At this point, there are infinitesimal networks which are more or less cross-linked to each other (b-stage). The gelation leads only to a small change in the reaction rate, but could be estimated with rheological measurement. The point of gelation is for a thermoset on a fixed degree of cure with the conclusion, that it is independent of the polymerization method (radiation, convection oven or microwave). Because of that, the measurement of the gelation point allows for calculating the activation energy of the polymerization reaction [7].

With the increase in the polymer network, the material becomes more and more rigid. The movement of the monomer reactants slows down abruptly when the glass transition point (T_g) of the polymer overwhelms the process temperature. The material vitrificates as a solid state, for polymers often called the glassy state, based on their more or less amorphous structure. The reaction rate drops down by an order of two or three magnitudes, and is significantly dependent on the mobility of the reactive monomers. In gel state, the polymerization is controlled by a chemical reaction rate, but in glassy formation, the reaction is controlled by the diffusion of the reactants. The glass transition temperature increases with the polymerization degree. Based on that, the glass transition temperature is highly dependent on the cure process. The vitrification is, in contrast to the gelation, reversible by heating. When the process temperature is again above the T_g devitrification occurs. A fully cured thermoset (c-stage) has a glass transition temperature near the dissociation temperature, which makes the measurement of such a T_g in most cases impossible. Based on that it can be postulated that a fully polymerized thermoset does not have glass transition temperature anymore [8].

3.2. The Polymerization Reaction Of Bcb

The fully polymerized BCB has a high 3-D cross-linked network, which leads to isotropic properties. The polymerization reaction consists of two steps. The first step is a thermal activated BCB four-membered ring opening reaction. This reactive intermediate readily undergoes a [2+4] Diels-Alder reaction (see figure 3). The Diels-Alder reaction is a reaction where a conjugated diene will be added on a double bond, building a six terms ring. Both reaction partners are named dien and dinophile. The reaction is also called a [4+2] cycloaddition.

After ring opening, the BCB monomer has two diens and dinophiles, which allow a highly cross-linked polymer. The ring opening is a first order reaction, and depends on the concentration of the pre-polymer. The reaction rate of the ring opening and the Diels-Alder reaction was measured by ROTH et. al. The ring opening has a 10 times higher reaction rate constant [9] [10].

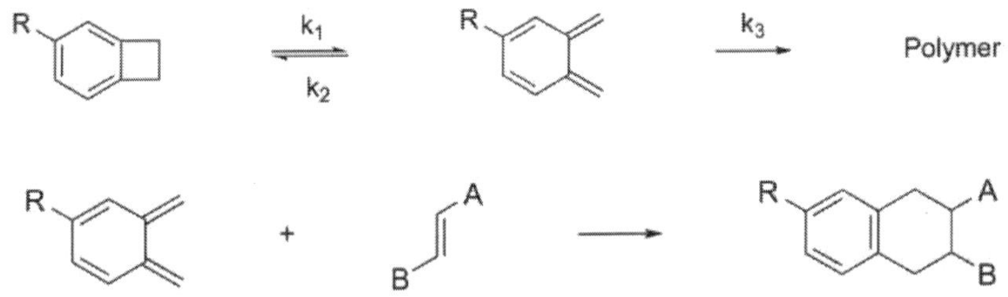

Figure 3. Ring opening reaction and Diels-Alder reaction of BCB

The polymerization reaction could be described by a pseudo first-order reaction equation, because the Diels-Alder reaction is the reaction rate limited factor. The thermal activated polymerization process of BCB or the reaction rate depends on temperature, time and extent of cure. Figure 4 show a Time-Temperature-Transformation diagram (TTT-diagram) for the polymerization reaction of BCB. The numbers in the diagram describes the extent of cure which could be measured by different methods.

The reactions rate is strongly influenced by the mobility of the molecule chains which generally increase with the temperature, but also decrease with the extent of cure, because increased cross-linking decreases the mobility. During the cure, the reaction kinetic runs through different stages, from liquid pre-polymer through a gelation to a vitrification.

The gelation point has no significant effect on the polymerization reaction. The change into the solid state impacts the reaction rate based on the change from a chemical controlled to a diffusion controlled regime. That leads to a significant drop in molecule mobility, which could decrease the reaction rate down to nearly zero. The slowdown factor of BCB is relatively weak because of the small monomers.

The commercially available CYCLOTENE (see table 1) has been pre-cured (b-stage resin). Especially for structured photo BCB layers, a viscous liquid state during processing is avoided by a higher molecular weight of the b-stage state. After the evaporation of the solvent, the polymer layer lost its flow ability and the layer could be structured. A flowing of polymer during the cure would limit the critical structure dimension. The reaction rate is nearly constant during the entire gelation phase. The fully cured BCB has a T_g near the decomposition temperature above 350°C. But the T_g of an uncured CYCLOTENE will exceed approximately 60°C, and will rise dependent upon the degree of cure (figure 5) [6].

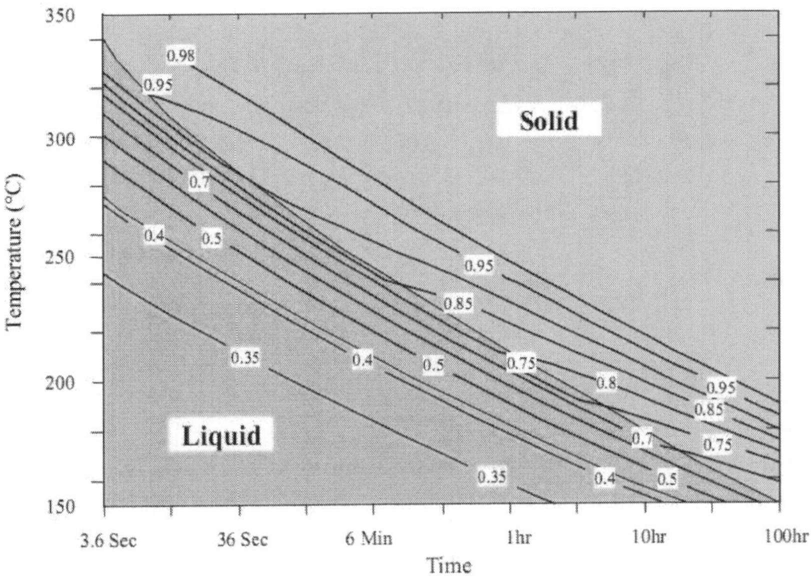

Figure 4. Extent of cure of photo BCB in relation to soak temperature and soak time. The colors describe the material conditions: liquid state (green), gel state (blue) and solid state (red) [**11**]

It is shown for some other polymers that the degree of cure is nearly fixed after vitrification without a temperature rise, because of the extremely slow reaction rate [12]. This means for polymerization process optimization a full cure is not possible if the temperature is decreased. A lower processing temperature leads to the glass transition temperature occurring earlier, which can be seen in figure 4. The polymerization process stops at a lower degree of cure, because the reaction changed earlier from a reaction controlled to diffusion controlled characteristic.

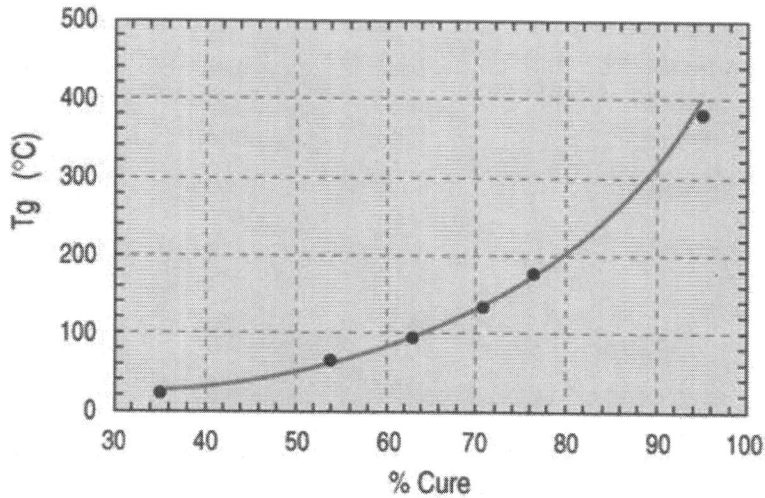

Figure 5. Glass transition temperature of photo BCB [**6**]

The slowdown of the reaction rate of BCB after vitrification has been described in previous publications [13] [14]. However, the reaction does not stop and proceeds to the solid state. In contrast to many other polymers, the diffusion controlled reaction rate is high enough to continue the polymerization process. DIBBS et. al. specifies only that above 95 % the diffusion controlled reaction rate will slow down so that it could be seen as a fixed state [2]. A degree of cure of about 95 % can be accepted as nearly fully cured. There is no significant difference measurable because of the approximation failures of measuring methods for the degree of cure.

4. MEASUREMENT OF THE POLYMERIZATION DEGREE OF BCB

The datasheet of CYCLOTENE describes a cure at 250°C for 90 minutes to get a full cured film. Any change in the cure process or the creation of a kinetic model needs knowledge about the extent of cure. There are a couple of methods to monitor it, like Fourier transformed infrared-spectroscopy (FT-IR spectroscopy), change of the refractive index or the difference scanning calorimetric (DSC).

The determination of the reaction rate, and the extent of reaction by time and temperature are investigated further to develop a kinetic reaction model. Real-time measurements during the cure process are necessary to monitor the degree of cure. DSC measurement was used for kinetic model creation. Based on the kinetic model, cure processes were set-up and the expected extent of cure was controlled by FT-IR spectroscopy. The FT-IR spectroscopy allows the measurement on product wafers, but for the creation of a kinetic model, a special oven with integrated measurement equipment is necessary. The discrete IR wavelength gets into resonance with the rotation or vibration of some molecule groups. The rate of reactants and products of the polymerization could be estimated by the peak height [15]. The samples were measured before and after the curing by the "FT-IR 2000 System" from Perkin-Elmer.

The broadly used method for the investigation of thermal processes like polymerization is difference scanning calorimetric. The DSC is preferred because of an accurate temperature regulation and good recording of the energy dissipation [12] [16]. The DSC-analysis was done with 7 mg CYCLOTENE 4024-40 in an inert atmosphere using a "Q2000" machine from the TA Instruments Company. The samples were dried and exposed before they were analyzed. The DSC analyzer could perform high heat-up and cool-down rates, up to 300°C/min. Higher heat-up rates leads to more inaccuracy based on a stronger overshoot of the thermal system.

The DSC records the heat flow to a sample during a tempering in comparison to a reference cell. The integration of the heat flow over time gives the heat dissipation of the sample. The polymerization reaction is an exothermic reaction and occurs in DSC plot by a dip/peak of the heat flow curve [14]. The integration of the heat flow peak over time allows the estimation of the heat transformation (enthalpy change) of the reaction.

The DSC experiments for the estimation of the BCB polymerization has been done in isothermal and non-isothermal mode with a constant heat rate. In figure 6 a non-isothermal signal of a BCB DSC sample is shown. The heat of the reaction of 294.4 J/g is calculated by a linear baseline. A more accurate calculation can be done by using a blind curve, which means a second measurement of the fully polymerized sample. The difference between the two curves gives the energy dissipation of the polymerization process. The blind curve also compensates the parasitic endothermic signal which came from heating up the aluminum sample pan [8].

The heat of the reaction could be set in relation to the extent of reaction. The reaction rate dα/dt at any state of the polymerization can be estimated at any stage of the reaction by the following relation:

$$\frac{d\alpha}{dt} = \frac{dH/dt}{\Delta H_{total}} \qquad (1)$$

The total heat of reaction ΔH_{total} describes the integral of heat flow from the initiation of reaction up until the full polymerized material form. The constant heat rate β allows for applying the extent of reaction over the heat rate, instead of the extent of reaction over time, which is measured with a higher inaccuracy:

$$\frac{d\alpha}{dT}\beta = \frac{dH/dt}{\Delta H_{total}} \qquad (2)$$

This is the reason why non-isothermal experiments are often used to analyze the reaction. The degree of cure at any state could be estimated by the rate of the heat of reaction ΔH to the total reaction heat ΔH_{total} of a fully cured layer:

$$\alpha = \frac{\Delta H}{\Delta H_{total}} + \alpha_i = \frac{1}{\Delta H_{total}} \int \frac{dH}{dt} dt \qquad (3)$$

where α_i is the initial degree of cure of the b-stage resin. The estimation has included an inaccuracy by the calculation of the total heat reaction of the complete polymerization. Linear heating experiments at different heating rates are performed. The ΔH_{total} of the non-isothermal experiments with different heat rates are printed in table 2. The blind curve corrected DSC-curves are shown in figure 7. By increasing the heat rate, the reaction is shifted more and more into higher temperatures. The average total heat of reaction of the fully cured BCB film is about 281.8 ± 5.3 J/g. The results are in good correlation to the measurement of CHAN, who measured a heat of reaction for CYCLOTENE 4024-40 of 291.3 ± 8.8 J/g [17]. The measurement of the heat reaction is strongly dependent on base line settings. The estimation of the total heat reaction has in general an inaccuracy of 3 to 5%. It should be noted that the photosensitive CYCLOTENE will be shipped out with an initial degree of cure of about 47 %. Based on this, the true total heat of the reaction is around 531.7 J/g. The value is confirmed by BAIR et. al. which shows a value of 515 ± 10 J/g for the complete polymerization of a monomer BCB resin [18]. T An increased heat rate leads to more intensive reaction peak but with less time resolution, which increases the inaccuracy for a kinetic model development.

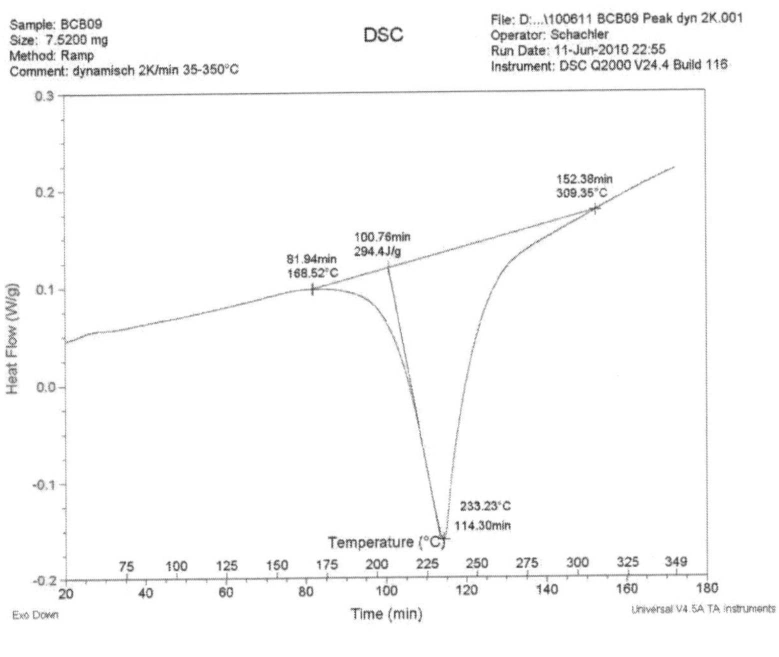

Figure 6. DSC plot of the polymerization of a photo BCB sample during a non-isothermal experiment with a heat rate of 2 K/min

Table 2. Heat reaction and the peak temperature of photo BCB

Heat rate [K/min]	Reaction heat ΔH [J/g]	T_{Peak} [°C]
2	289,6	233,35
5	283,6	245,24
10	276,4	255,52
20	281,8	266,00
40	277,4	277,12
Average ΔH_{total}	281,8 ± 5,3	

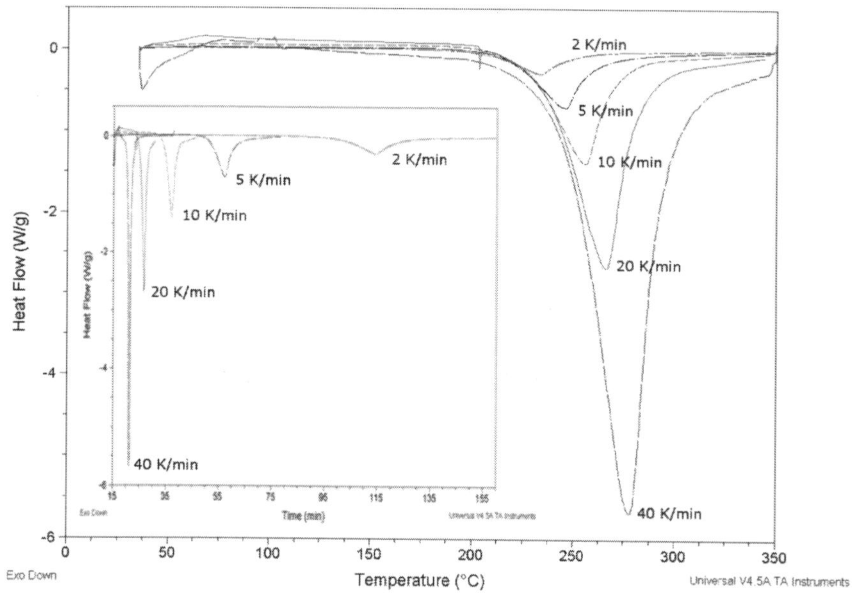

Figure 7. Non-isothermal DSC measurement of CYCLOTENE 4026-46 versus temperature and time

The state change at the glass transition temperature caused more or less a peak in the DSC curve. This peak increases significantly with higher heat rates [19]. The uncured coated BCB has a T_g of around 60°C. In figure 7 the parasitic peak which is caused by the overwhelming of the glass transition temperature (gelation) could be seen for the two heat rates at 20K/min and 40K/min in the temperature range from 60°C up to 100°C.

The vitrification of the polymer is the important factor for the reaction kinetic caused by the gradual decreasing of the rate. When the material returns to a glassy state, the same peak is generated, but the high slope of the reaction signal absorbs the additional peak. The vitrification could only be observed by the gradual change of the reaction rate.

The gradual change is in the case of non-isothermal experiments softened by the permanent increase in the temperature. The heat signal of the state change is calculated into the total heat of reaction.

Based on the knowledge of the T_g (figure 5), the influence of the vitrification could be estimated for the different heat rate experiments. The theoretical development of the T_g for the non-isothermal experiments is plotted in figure 8. The influence of the diffusion controlled part decreases with higher heat rates and has shifted the vitrification to higher temperatures.

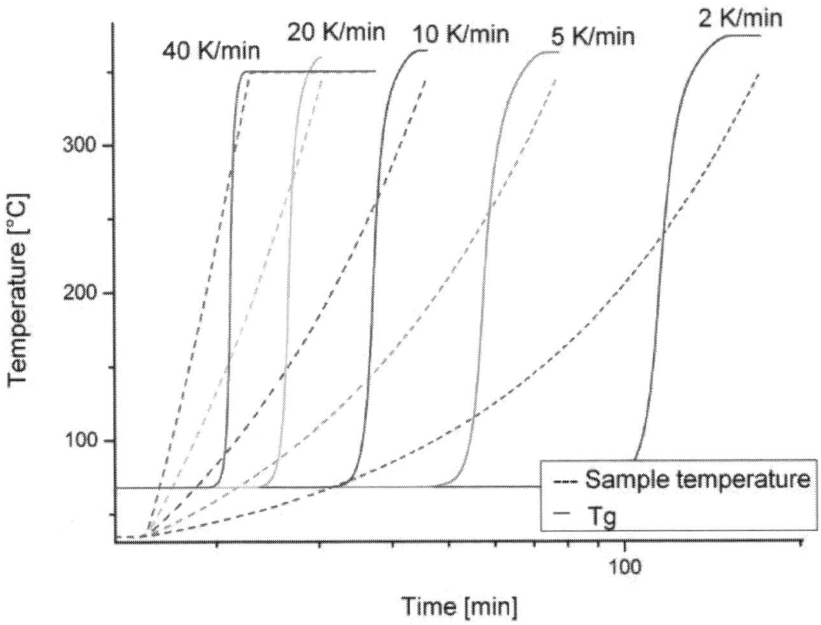

Figure 8. Prediction of the glass transition temperature for the non-isothermal experiments of photo BCB

Low heat rates or isothermal experiments allow to analyze better the influence of the virtification on the reaction rate. An isothermal measurement is based on a heat up phase, a soak temperature and a cool down phase. An isothermal experiment also gives a better prediction for the real cure process in a convection oven. But an isothermal measurement method has some disadvantages in contrast to an experiment with a continuous heat rate. The extent of cure is calculated by the integration of the reaction peak signal to get the reaction heat ΔH_{ds} over the soak temperature. The higher time resolution of the reaction signal is in relation to a much smaller reaction peak. The small peak and the possibility of not a fully polymerized layer at the soak temperature, caused by vitrification or short process time, are critical impacts in estimating the total heat reaction ΔH_{total}. A simple estimation of the total heat reaction can lead to an inaccurate interpretation of cure degree (see Eq.(3)). The reaction already starts during the heat up and cool down phases, which takes into account the integration of heat flow over time. The discontinuity of the temperature function produce parasitic signals. The DSC signal of a BCB soft cure at 210°C at 30 minutes is plotted in figure 9. The heat flow signal shows that there is a parasitic signal peak between isothermal and dynamic heat phases. The reaction start during the heat up phase, which can also be seen. The short process time leads only to a soft cured layer, which means that the measured reaction heat does not equates to the total reaction heat. After the soft cure, the sample was again heated up to 350°C with a constant heat rate. A leftover heat of reaction ΔH_{res} can be measured. The sum of ΔH_{ds} and ΔH_{res} leads to the total heat of reaction ΔH_{total} [12].

5. DEVELOPMENT OF A KINETIC MODEL OF THE POLYMERIZATION REACTION

The extent of cure for a dedicated cure program can be measured by DSC experiments. But the prediction of the extent of cure and the polymerization behavior based on any temperature – time program is necessary for any process optimization with respect to energy consumption, process time or material properties. On account of the DSC experiments, a kinetic model can be set up to characterize the polymerization reaction and can be used to design and optimize the cure process. The reaction is described by a mathematical relation between time, temperature and the extent of cure. In general, special software is required to develop the kinetic equation. The kinetic model in combination with a chemical-rheological model allows a simulation of the complete polymerization reaction [14] [8].

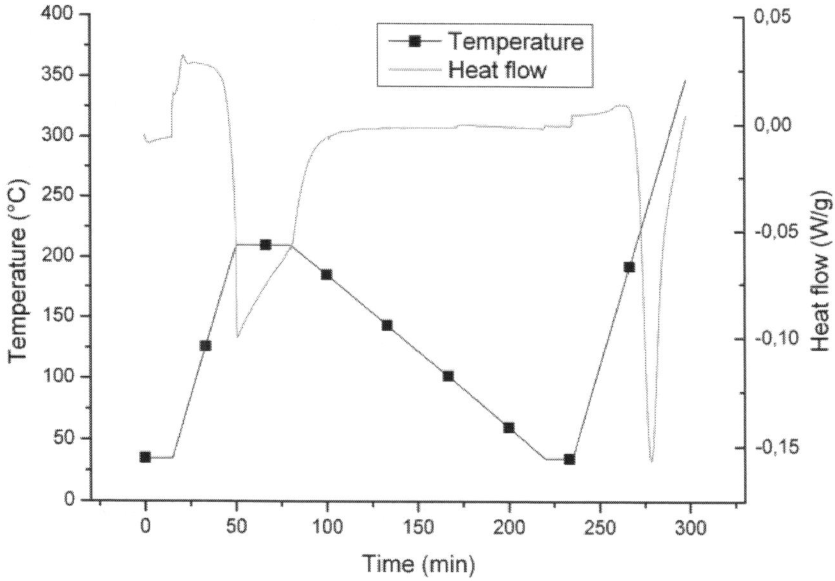

Figure 9. DSC measurement of a BCB soft cure process with additional non-isothermal heat up step to estimate the cure degree.

The kinetic models are generally sorted into two types:

- A phenomenological model and
- A chemical model.

The phenomenological model describes the reaction by a relative simple equation. This model type intentionally ignores details of the reaction with the benefit of simple application. Detailed relations and progression of the different chemical reaction steps and the relationship of the reaction groups will be ignored. The chemical reaction model takes into account the chemical steady state of the partial reactions. These types of models have a much better prediction and interpretation potential, but it is often not useable, especially for complex reactions. In contrast to a phenomenological model, the efforts involved for such a model are much higher, because many more reactions parameters are necessary, and the users need to do a chemical analysis of every reaction step. The phenomenological model can be used without chemical understanding, and can be created only by a comparison of reactants and products. The extent of cure behavior, which has been estimated by DSC experiments, was used to create a phenomenological model for BCB. The simple application makes the phenomenological model the most preferred for polymerization reactions [20] [21].

The two-step polymerization reaction of BCB can be described by a pseudo first-order equation. Based on that simplification, the BCB reaction could described as follows:

$$\frac{d\alpha}{dt} = k_e(T)f(\alpha) \tag{4}$$

Where α is the degree of cure, $k_e(T)$ is the reaction rate constant which depends on the temperature, and $f(\alpha)$ is the reaction model which describes the reaction order [22]. The kinetic equation such as Eq.(4) is a simplified assumption which describes the reaction rate as a product of a temperature-depending function and a fraction-transformed-depending function. There is a range of empirical mathematical forms for the reaction model [23]. Some particularly idealized reaction models are proposed in table 3.

The reaction rate constant could be realized for the simplest case by the Arrhenius relation:

$$k_e = A\exp\left(\frac{-E_A}{RT}\right) \tag{5}$$

Where R is the universal gas constant, E_A is the activation energy, T is the temperature and A is the reaction rate constant at an infinite temperature [24]. The Arrhenius equation describes a reaction in gas phase. In approximation, the relation can be used for liquid and solid materials. But influence by molecule mobility and vitrification are ignored and could be added by the adaptation of the reaction function.

Table 3. A range of reaction models which are often used in kinetic analytics

Reaction model	f(α)
Power law	$4\alpha^{3/4}$
Power law	$3\alpha^{2/3}$
One-dimensional diffusion	$1/2\alpha^{-1}$
Mampel (first order)	$1-\alpha$
Avrami-Erofeev	$4(1-\alpha)[-\ln(1-\alpha)]^{3/4}$
Three dimensional diffusion	$2(1-\alpha)^{2/3}[-\ln(1-\alpha)^{1/3}]^{-1}$
Contracting sphere	$3(1-\alpha)^{2/3}$

The determination of the three important kinetic parameters, the two Arrhenius parameter E_A and A and the reaction model f(α) is an interlinking problem. The measurement of one of the parameters, and especially the accurate estimation, influence the other two parameters. Based on this problem they are often called a kinetic triplet [25]. There are many different methods for the determination based on DSC measurements. The method of KISSINGER has been used for the investigation of the polymerization process of BCB. It allows for a simple application with a small inaccuracy, which is also described in detail by STARINK.

The KISSINGER method is focusing on the determination of activation energy as first parameter with the most accuracy possible. The reaction rate in Eq.(4) is calculated by the Eq.(1) from a DSC experiment. The Eq.(5) is inserted into the Eq.(4). The logarithm of Eq.(4) leads to the following equation:

$$\ln\frac{d\alpha}{dt}=-\frac{E}{RT}-\ln f(\alpha) \qquad (6)$$

For non-isothermal experiments with constant heat rate (β=dT/dt) the Eq.(6) could be changed to:

$$\ln \frac{d\alpha}{dT}\beta = -\frac{E}{RT} - \ln f(\alpha)\,. \tag{7}$$

The activation energy could be obtained from the slope of plots of ln(dα/dT) versus 1/T. It is estimated without any knowledge of the reaction model f(α). In literature, the principle is also termed model free kinetic (MFK). The slope in the plot should be constant. If there is a change in the slope, it indicates a phase change in the reaction. The calculation of the reaction rate by Eq.(1) leads to an inaccuracy which could have a huge impact on the kinetic equation caused by the triplet relation. The assumption of the reaction rate can be replaced by correlation of a set of non-isothermal DSC experiments. Based on Kissinger, the Eq.(7) is integrated by separation of variables and the following approach was postulated:

$$\int_0^{\alpha_f} \frac{d\alpha}{f(\alpha_f)} = \frac{A}{\beta}\int_0^{T_{100\%}} \exp\left(-\frac{E_A}{RT}\right)dT \approx \frac{AE_A \exp\left(\frac{-E_A}{RT}\right)}{\beta R \cdot \exp\left(\frac{-E_A}{RT}\right)^{\kappa}} \tag{8}$$

The Eq.(5) have been logarithm and lead to following equation:

$$\ln \frac{\beta}{T_f^{\kappa}} = -\frac{E}{RT} + C \tag{9}$$

C is a constant for different heating rates at a constant degree of polymerization and belongs to the reaction model. T_f means the temperature of experiments with different heat rates where the same extent of cure occurs. The variable κ is a correction factor in the range of 1,92 up to 2. With the Eq.(9)the E_A can determined as the slope of the plotln(β/Tκfln□(β/Tfκ) versus 1/TfTf independent of constant C. Three or more experiments with different heating rate β are necessary to determine the activation energy with MFK. The correct estimation of T_f is the main challenge at this method. For the normal reaction, analyzing the maximum rate method or peak method is used, which is described also in the ASTM E698 norm [26]. The peak method avoids the defective calculation of cure rate by Eq.(1). It was found that the extent of cure is approximately fixed at the maximum reaction rate, but a small error will be introduced. The peak temperatures of the non-isothermal experiment are shown in table 2. The temperature of the peak is shifted with the heat rate. The results of Eq.(9) for the five non-isothermal experiments are plotted in figure 10.

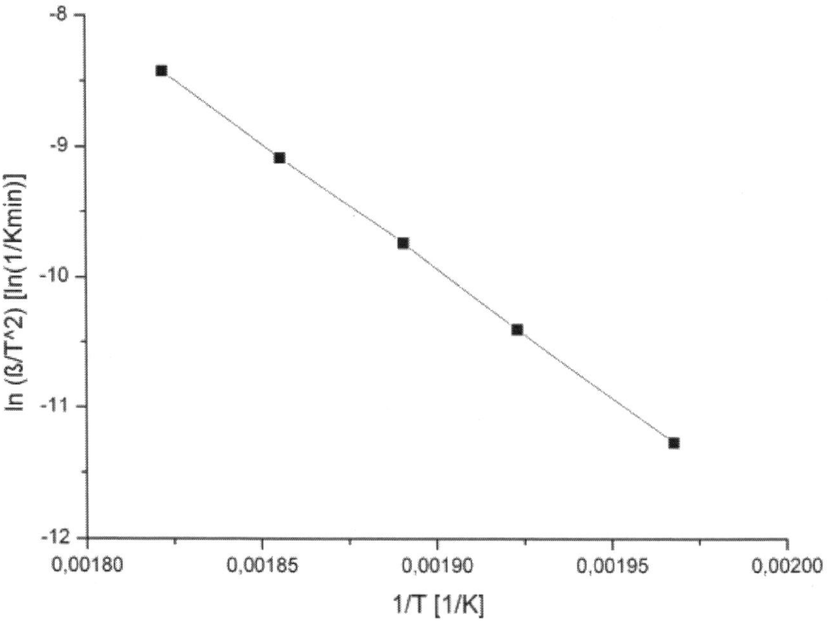

Figure 10. Non-isothermal measurement of BCB. The slope approximate the activation energy of the polymerization reaction cure degree of around 73%

A value of 149.5kJ/mol ±1.75 kJ/mol was measured, which acknowledges the declarations of DIBBS et. al.. Different literature sources describe an activation energy in a range from 146 kJ/mol up to 197,6 kJ/mol [27] [17] [28]. A disadvantage of this method is that the values are only estimated at one degree of cure (around 73 % conversion). Therefore, in relation to the used heat rates, the value describes only the chemically controlled kinetic. With a first-order reaction model, the A has a value of 1,35 h^{-1} for a reference temperature T_0 of 210°C:

$$\frac{d\alpha}{dt} = 1{,}35h^{-1}\exp\left[-\frac{149{,}5kJ/mol}{T}\left(\frac{1}{T}-\frac{1}{T_0}\right)\right](1-\alpha) \tag{10}$$

The plot for the non-isothermal experiments (Figure 11) shows that the model fits the chemically controlled regime very well. The estimated model is in good agreement with the already proposed kinetic model for BCB [2] [6]. There is a poor estimation for the diffusion controlled part, which is also described in previous publications [6] [28]. With the focus on lower cure temperatures, the influence of the diffusion controlled part increases.

The polymerization analysis aims for a better understanding of the influence of the vitrification. That demands a modulation of the normal kinetic model Eq.(10). There are many different methods for correcting the conventional reaction model by a conversion-dependent diffusion contribution. One of the most popular methods for thermosetting systems is to modify the kinetic equation by a "diffusion control function" which is described by SCHAWE [29].

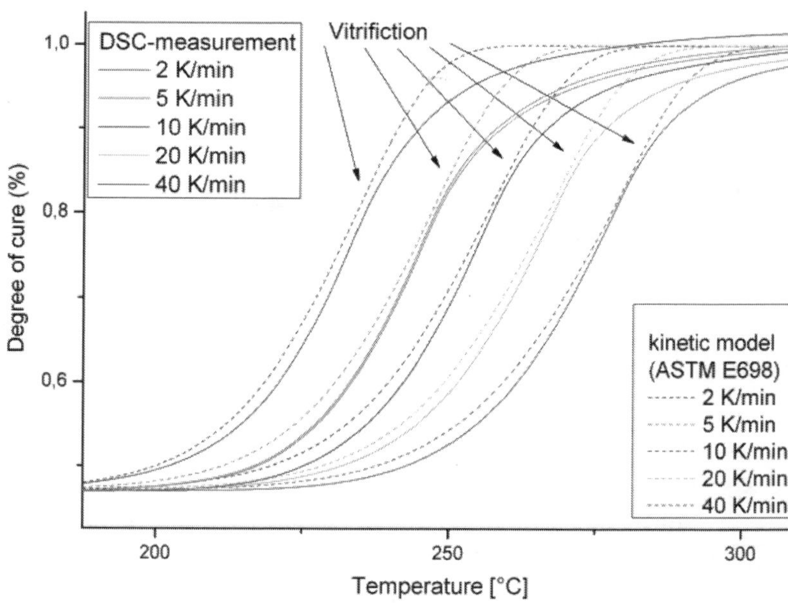

Figure 11. Extent of polymerization based on the non-isothermal experiments and calculation based on Eq.(10)

$$\frac{d\alpha}{dt} = k_e(T) f(\alpha) f_d(\alpha;T) \tag{11}$$

The diffusion control function works in an electro-technical analogy like a low-pass filter. During the polymerization up to the change in the solid state, the function should not influence the reaction which means $f_d(T<T_g)=1$. After the vitrification, the function decreases down to zero. A function which achieves this condition is proposed by SCHAWE:

$$f_d(\alpha;T) = f_d(T_g(\alpha)) = 1 - \left(1 + \left(\frac{T_0 - T_g(\alpha)}{D}\right)^3\right)^{-1} \tag{12}$$

where T_0 is the temperature at which the diffusion control function is zero and A is an empirical constant to smooth the function. The function is focused for a reaction which is nearly stopped in a glassy state. The difficulty of this method is to estimate the empirical diffusion control function which leads to the best fit. Another problem is the influence of the diffusion control function for the chemically controlled part.

MENCZEL describe an alternative model from RABINOWITCH for small-molecule reactions, where the rate constant k is based on a sum of a chemically controlled and diffusion controlled rate constant [8]:

$$\frac{1}{k_e(T,\alpha)} = \frac{1}{k_r(T)} + \frac{1}{k_d(T,\alpha)} \tag{13}$$

where k_e is the overall rate constant, k_r is the Arrhenius ratio constant for the chemical reaction and $k_d$is the diffusion rate constant. The overall reaction rate constant k_e is governed prior the vitrification by the Arrhenius rate constant, and after vitrification is dominated by the diffusion rate constant. The disadvantage is an extended interlink problem of the parameters.

A combination of both models from SCHAWE and RABINOWITCH has been used to create an advanced model for the polymerization reaction of CYCLOTENE. Two different rate constants (k_r;k_d) were calculated. The overall reaction rate constant is the sum of both, but the terms are combined with a step-function, which works like a switch and allows the change between gel and glassy state kinetic models.

$$\frac{d\alpha}{dt} = k_r(T) f(\alpha) f_{step}(\alpha;T) + k_d(T) f(\alpha)(1 - f_{step}(\alpha;T)) \tag{14}$$

The step-function belongs to the process temperature and the glass transition temperature and gets a value between one and zero. For both reaction regimes, a first-order reaction model was used based on a good fit, but the model allows for the declaration of two different reaction models for chemically controlled and diffusion controlled states. The following expression was used for the step-function:

$$f_{step}(\alpha;T) = \frac{1}{1 + \exp(-(T - T_g(\alpha))B)} \tag{15}$$

where B is a parameter to smooth the changeover between the two reaction functions. The value was set at 0.1 to get a change in temperature range of ±5°C around the glass transition temperature. The glass transition temperature depends on the extent of cure (see figure 5), and the following relation was estimated based on the measurements of TOEPPER:

$$T_g = \frac{35}{0.054}\alpha^2 - \frac{10865}{27}\alpha + \frac{3070}{27} \tag{16}$$

For the kinetic model, two Arrhenius terms need to be calculated, one for the chemically controlled part and one for the diffusion controlled part. The Eq.(9) allows the calculation of the activation energy for different extents of reaction. In contrast to the peak method, the extent of reaction need to be calculated by Eq.(3). The consideration of the vitrification-heat rate relation for the estimation of the activation energy of the two regimes is necessary to see any difference. For example, the highest heat rate of 40 K/min leads to a vitrification above 90% in contrast to the slowest heat rate of 2 K/min, which vitrificates around 80%. Therefore the heat rate of 2 K/min is used to calculate the activation energy for the chemically controlled state up to 80%, above this value it is used for the calculation of the activation energy of the diffusion controlled state. The calculated activation energies for different extents of reaction are shown in figure 12.

The change in the reaction kinetic is marked by the change in the activation energy. The activation energy for the chemically controlled region has an approximate value of 157.1 ± 3.6 kJ/mol. The activation energy is slightly higher than the value which was estimated by the peak method /ASTM E698 norm (pointed line). For the diffusion controlled kinetic, a higher activation energy of around 166.1 ± 8.1 kJ/mol was measured. The increase in the activation energy in a diffusion controlled regime shows a higher temperature dependency for the reaction, which could be explained by the more temperature-dependent molecule mobility.

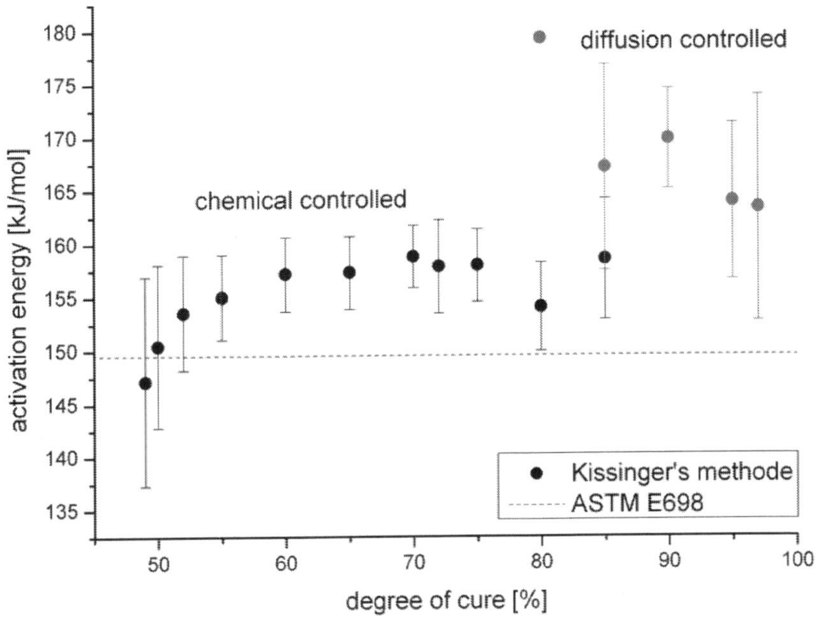

Figure 12. Activation energy estimated by the ASTM E698 norm and dependent of the degree of cure

The results for the reaction rate at the reference temperature are 1.41 hr^{-1} for the chemically controlled regime, and 0.27 h^{-1} for the diffusion controlled regime. In figure 13, the calculated degree of polymerization for the conventional (Eq.(10)) and modified kinetic model (Eq.14) is compared to a measured curve for a cure at 190°C. The modified model shows a much better fit in comparision with the real behavior above a degree of cure of approximately 80 %.

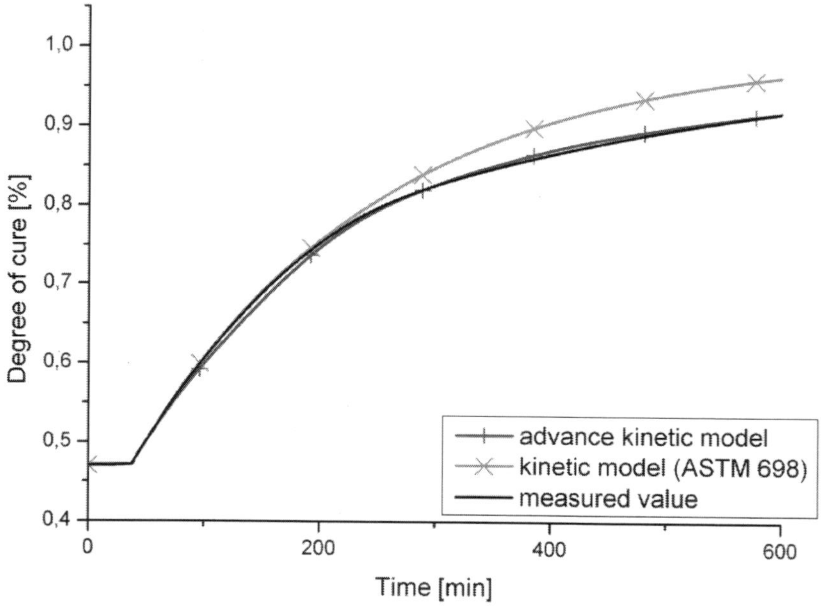

Figure 13. Degree of cure of a photo BCB sample at 190°C

6. BENEFITS OF THE KINETIC MODEL FOR THE BCB POLYMERIZATION AND THE RELATION FOR THE POLYMER PROPERTIES

The investigation of the cure behavior of BCB shows that an ongoing polymerization reaction in solid state is also possible with the requirement of a minimal process temperature above 150°C caused by the thermal activation. The enhanced model which is described in Eq.(14) allows a much better prediction for the polymerization of BCB in glassy state. The polymerization in vitrificated state is relative slow and should only use if the device is temperature critical. But the cure in glassy state could be interesting with the focus on the properties of the polymer. The stress development of BCB during the cure at 250°C is plotted in figure 14. The thermal stress of a polymer film on a substrate could be calculated as followed:

$$\sigma_{th} = \frac{E_f}{1-\upsilon_f}(\alpha_f - \alpha_s)\Delta T \tag{17}$$

Where E_f and υ_f are the Young's modulus and Poisson ratio of polymer. For silicon as substrate CTE α_s are approximately 2.6 µm/mK, and BCB has a CTE of 42 µm/mK. Delta T is the temperature difference between ambient temperature and the vitrificated temperature. The polymer is in gel phase during the heat up phase and the stress is nearly zero. When the polymer vitrificates, the stress state at that point is set to zero. During the cool down, the thermal tensile stress in the layer increases based on the coefficient of thermal expansion mismatch to the substrate [8] [30] [31]. A nearly fully cured BCB has a T_g near the decomposition temperature, which means the zero stress point could not reset by processing at a higher temperature. A reheating of the BCB after the cure at 250°C above the temperature where the polymer vitrificate leads to a compressive stress, which could be seen in figure 15. The stress measurements were done by Toho Technology with a Flexus-2320-S system. In comparing the stress values at room temperature between the measurement of figure 14 and figure 15 a slightly difference occurs. This could be explained by a visco elastic behavior of the BCB which is discussed elsewhere [32].

There are trends in microelectronics towards enlarging the wafer size in combination with thinner wafers to decrease costs. In contrast to this, the signal frequency gets higher and higher, which means packaging thick dielectric polymer layers to assure good signal integrity. It results in two major challenges. The first is in handling, because the stress leads to a bow of the wafer, which is a problem for automatic handling tools and also for the processes because of the topography. The second point is that the stress in the layer stack is a continual problem for reliability. It force cracks through the layers and delamination. The decrease of the stress in the layer stack becomes more and more a focus in the future. This could also be seen by the polymer supplier who presented a couple of new polymers with low temperature cure properties in the last few years.

The stress in the polymer layer can be minimized by the optimization of the cure process with the help of a kinetic model. There are already some works which try to decrease the stress by cure process optimization [33] [34], but they unaccounted the meaning of the vitrification for the stress formation. A set of BCB experiments was done to examine the potential of stress decreasing by process optimization. In relation to the lowest temperature where vitrification occurs, and a nearly fully cured layer is obtainable, a theoretical minimal stress of around 18 MPa should be achievable. There has been shown a possible decreasing of the stress in a BCB layer from 28 MPa based on a normal cure at 250°C down to a stress value of 19 MPa [32]. Nevertheless, the process time will increase exponentially with a decrease of the temperature, and for higher volumes a decrease of the temperature down to 150°C is not economically feasible. The kinetic model allows for the creation of curing processes with more than one soak temperature, with the benefit that the processing temperature before and after the vitrification is higher to decrease the processing time, and the temperature is set lower in the phase change to a solid state. The cure of the polymer is generally performed in convection ovens. The slow temperature change in a convection oven limits the difference between the high and low temperature levels. A programmable control of the oven makes a dynamic heat process possible, which reaches the shortest possible time in combination with low stress in the polymer layer. Such a theoretical temperature profile is plotted infigure 16.

The oven will heat up to a high peak temperature to achieve fast polymerization. After that the temperature is decreased to reach a vitrification at low temperature. The temperature will be ramped up after the vitrification, but the process temperature is controlled to be below the glass transition temperature, to decrease the process time and avoid devitrification. The described program aims for a vitrification at low temperature in relation to the stress in the polymer layer. DIBBS described a seven days process, which also aims to get a vitrification at a low temperature and slowly increase the temperature after that [2]. The process involves a long time processing at 150°C, and after that a slow ramp up to 250°C. The expected cure profile of DIBBS process is plotted in figure 17. The long heat up phase of 6 days results due the lack of a detailed kinetic model for the diffusion controlled part.

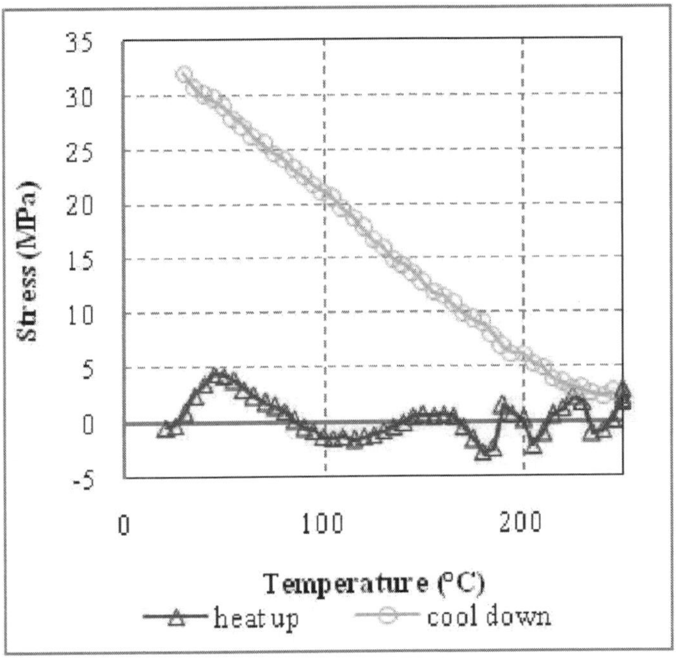

Figure 14. The stress in the material during a cure at 250°C

He realized only a small decrease of the stress down to 24 MPa. The results of DIBBS and own experiments lead to the assumption that the ongoing of the polymerization in glassy state also influences the stress. A BCB layer on substrate was soft cured at 210°C up to 70%, which has a T_g of around 190°C. After soft cure, the layer was stored at 150°C for 100 hours and reached 93% polymerization. The stress in the layer decreased during the storage from 24 MPa down to 19 MPa [32]. A layer which was fully cured at 250°C shows no strong relaxation after the storage. This could be explained by two things. First, the partly cured film has a higher viscoplastic behavior and the material relaxed easier.

Another possible reason is stress relaxation caused by the ongoing polymerization at low temperature. Further ongoing experiments are necessary to clarify in more detail the stress behavior in solid state BCB layers.

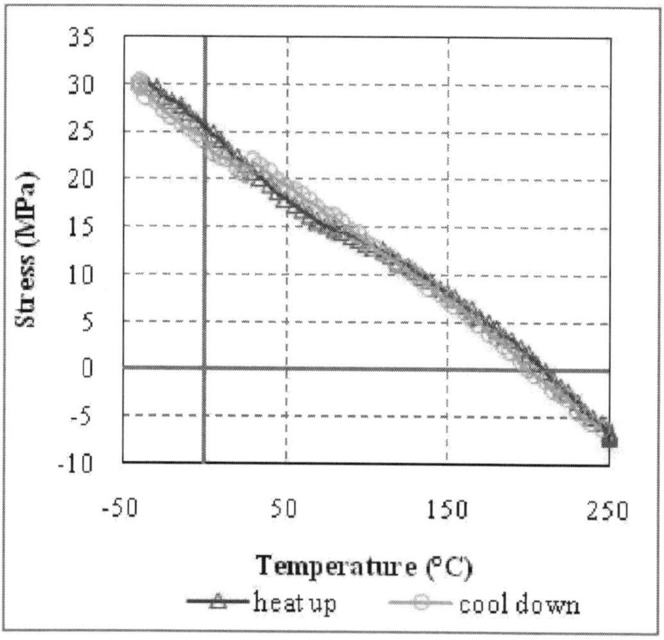

Figure 15. Stress curve of a cured BCB layer on a silicon substrate measured with Flexus-2320-S

The polymer dielectric layer is an essential component in packaging technology. The temperature-activated polymerization processing of polymers has increasingly become a key parameter for the process workflow. The investigation of polymerization processes allows the optimization of the cure process down to lower temperatures. Based on the example of BCB, a simple model was developed, which also described the polymerization reaction of BCB in a solid state. The model allows for significantly decreasing the stress in the layer and also decreasing down to very low cure temperatures. The mechanical properties like Young's modulus or tensile strength of BCB will not be affected by the processing parameter changes [32]. In combination with the low-temperature curing possibilities of BCB, a decrease in the stress of about approximately 30% is possible.

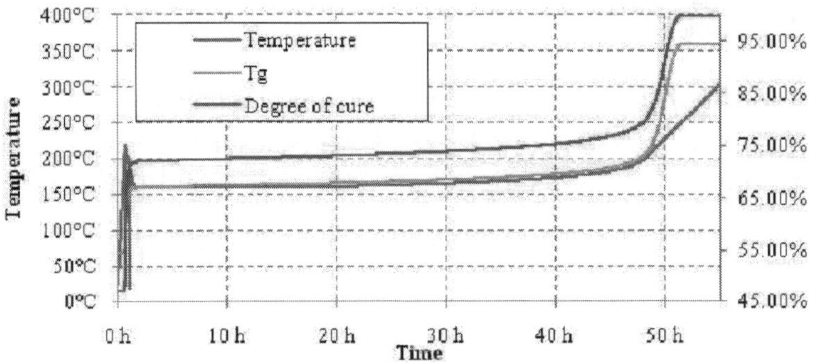

Figure 16. Calculated process profile with a focus on a vitrification at low temperature

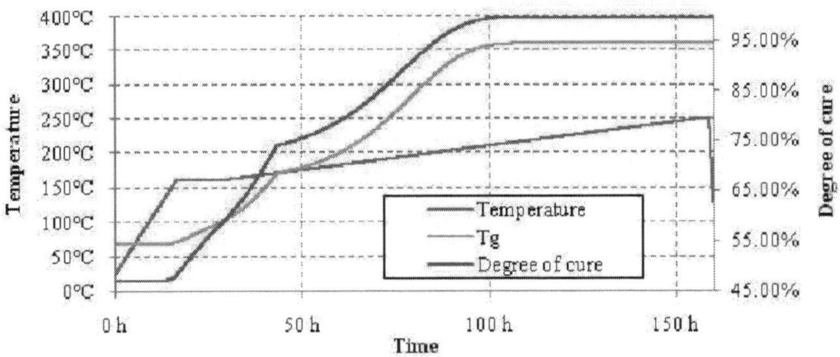

Figure 17. Calculated process profile of the low stress cure of DIBBS

ACKNOWLEDGEMENT

The authors want to emphasize the contributions from the colleagues at Fraunhofer IZM who are not mentioned here by name. Financial support from the Deutsche Forschungsgemeinschaft (International Research

Training Group GRK 1215, "Materials and Concepts for Advanced Interconnects") is gratefully acknowledged.

REFERENCES

1. M. Toepper, T. Fischer, T. Baumgartner und H. Reichl, „A comparison of thin film polymers for Wafer Level Packaging," in Electronic Components and Technology Conference (ECTC), Proceedings 60th, Las Vegas, NV, USA, 2010.

2. M. Dibbs, P. Townsend, T. Stokich, B. Huber, C. Mohler, R. Heistand, P. Garrou, G. Adema, M. Berry und I. Turlik, „Cure Management of Benzocyclobutene Dielectric for Electronic Applications," in International SAMPE Electronic Materials and Processes Conference, Proceedings 6th, Baltimore, MD, USA, 1992.

3. N. Ghalichechian, A. Modafe, R. Ghodssi, P. Lazzeri, V. Micheli und M. Anderle, „Integration of benzocyclobutene polymers and silicon micromachined structures using anisotropic wet etching," J. Vac. Sci. Technol. B, Bd. 22, pp. 2439-2447, 2004.

4. R. L. Hubbard und R. S. Garard, „Wafer Level Curing of Polymer Dielectrics with Variable Frequency Microwaves," in International Conference on Advanced Thermal Processing of Semiconductors RTP, Proc. 15th, Cannizzaro, Catania, Italy, 2007.

5. G. Roelkens, J. Brouckaert, D. V. Thourhout, R. Baets und R. N. and, „Adhesive Bonding of InP/InGaAsP Dies to Processed Silicon-On-Insulator Wafers using DVS-bis-Benzocyclobutene," Journal of The Electrochemical Society, Bd. 153, pp. 1015-1019, 2006.

6. M. Toepper, Entwicklung einer auf Photo-BCB basierenden Technologie für das Waferlevel Packaging, Phd thesis, Technische Universität Berlin, 2004.

7. M. E. De, M. Moursb und H. H. Winter, „The Gel Point as Reference State: A Simple Kinetic Model for Crosslinking Polybutadiene via Hydrosilation," Polymer Gels and Networks, Bd. 5, pp. 69-94, 1997.

8. J. D. Menczel und B. Prime, Thermal analysis of polymers: fundamentals and applications, John Wiley and Sons, 2009.

9. W. R. Roth, M. Biermann, H. Dekker, R. Jochems, C. Mosselman und H. Hermann, Das Energieprofil des ortho-Chinodimethan-Benzocyclobuten-Gleichgewichtes, Chem. Bereichte 111, pp. 3892-3903, 1978.

10. W. R. Roth, T. Ebbrecht und A. Beitat, Chinodimethan-Benzocyclobuten-Gleichgewicht; Sauerstoffabfangtechnik, Chem. Berichte 121, pp. 1357-1358, 1988.

11. J. Yang, Processing Procedures for CYCLOTENE 3000 Series Dry Etch Resins, DOW Chemical Co., 2005.

12. H. Schindlbauer, G. Henkel, J. Weiss und W. Eichberger, „Quantitative Untersuchungen zum Aushärtungsverhalten von Phenoplasten mittels differentialthermoanalytischer Messungen (DSC)," Die Angewandte. Makromolekulare Chemie, Bd. 49, Nr. 725, pp. 115-128, 1976.

13. H. E. Kissinger, „Reaction Kinetics in Differential Thermal Analysis," Analytical Chemistry, Bd. 29, Nr. 11, pp. 1702-1706, Nov. 1957.

14. G. Widmann und R. Riesen, Thermoanalyse, Hüthig, 1990.

15. T. Stokich, W. M. Lee und R. A. Peters, „Real-Time FT-IR studies of the reaction kinetics for the polymerization of Divinyl Siloxane bis-Benzocyclobutene monomers," in MRS Proceedings, 227, 1991.

16. J. Dupuy, E. Leroy und A. Maazouz, „Determination of Activation Energy and Preexponential Factor of Thermoset Reaction Kinetics Using Differential Scanning Calorimetry in Scanning Mode: Influence of Baseline Shape on Different Calculation Methods," Journal of Applied Polymer Science, Bd. 78, pp. 2262-2271, 2000.

17. K. Chan, M. Teo und Z. Zhong, „Characterization of low-k benzocyclobutene dielectric thin film," Microelectronics International, Bd. 20, pp. 11-22, 2003.

18. H. E. Bair und C. A. Pryde, „Curing and glass transition behavior of a benzocyclobutene polymer," in Society Plastics Engineers Proceedings, Montreal, Quebec, Canada, 1991.

19. R. B. Prime, „An Introduction to thermosets," 2005.

20. V. M. Gonzales-Romero und N. Casillas, „Isothermal and Temperature Programmed Kinetic Studies of Thermosets," Polymer Engineering and Science, Bd. 29, Nr. 5, pp. 295-301, 1989.

21. A. Yousefi, P. G. Lafleur und R. Gauvin, „Kinetic studies of thermoset cure reactions," Polymer Composites, Bd. 18, Nr. 2, pp. 157-168, 1997. S. White und H. Hahn, „Process Modeling of Composite Materials: Residual Stress Development during Cure. Part I Model Formulation," Journal of Composite Materials, Bd. 26, pp. 2402-2422, 1992.

22. S. Vyazovkin, „Modell-free kinetics," Journal of Thermal Analysis and Calorimetry, Bd. 83, pp. 41-51, 2006.

23. K. J. Laidler, Chemical Kinetics, Harper & Row, 1987.

24. M. J. Starink, „The determination of activation energy from linear heating rate experiments: a comparison of the accuracy of isoconversion methods," Thermochimica Acta, Bd. 404, pp. 163-176, 2003.

25. B. Battenstein und H. Ritter, „Composite Membranen. Einfluß der Glastemperatur und der Zusammensetzung von Acrylsaure-Methylmethacrylat Copolymeren auf das Permeationsverhalten von Decan," Makromol. Chem., Bd. 7, pp. 767-770, 1986.

26. J. Kawahara, A. Nakano, K. Kinoshita, Y. Harada, M. Tagami, M. Tada und Y. Hayashi, „Highly thermal-stable, plasma-polymerized BCB polymer film," Plasma Sources Sci. Technol., Bd. 12, pp. 80-88, 2003.

27. P. E. Garrou, R. H. Heistand, M. G. Dibbs, T. A. Manial, C. E. Mohler, T. M. Stokich, P. H. Townsend, G. M. Adema, M. J. Berry und I. Turlik, „Rapid thermal curing of BCB dielectric," IEEE Transactions on Components, Hybrids, and Manufacturing Technology, Bd. 16, Nr. 1, pp. 46-52, 1993.

28. J. E. K. Schawe, „A description of chemical and diffusion control in isothermal kinetics of cure kinetics," Thermochimica Acta 388, pp. 299-312, 2002.

29. T. Hodge, N. Grove, B. Sinno, S. Bidstrup-Allen und P. Kohl, „Thermo-Mechanical Stresses of Spin-Cast Polymer Films on Silicon Substrates," The International Journal of Microcircuits and Electronic Packaging, Bd. 20, Nr. 1, pp. 12-20, 1997.

30. A. M. Bowles, „Stress Evolution in Thin Film Polymer," Phd thesis, Harvard University, 2006.

31. M. Woehrmann, M. Toepper, H. Walter und K.-D. Land, „Low temperature cure of BCB and the influence on the mechanical stress," in Electronic Components and Technology Conference (ECTC), Proceedings 61th, Orlando, FL, USA, 2011.

32. A. K. Gopal, S. Adali und V. E. Verijenko, „Optimal temperature profiles for minimum residual stress in the cure process of polymer composites," Composite Structures, Bd. 48, pp. 99-106, 2000.

33. S. White und H. Hahn, „Cure cycle optimization for the Reduction of Processing-Induced Residual Stresses in Composite Materials," Journal of Composite Materials, Bd. 27, pp. 1352-1378, 1993.

34. S. Vyazovkin und W. Linert, „False isokinetic relationships found in the nonisothermal decomposition of solids," Chemical Physics, Bd. 193, pp. 109-118, 1995.

35. S. Vyazovkin, „An approach to the solution of the inverse kinetic problem in the case of complex processes. Part 4. Chemical reaction complicated by diffusion," Thermochimica Acta, Bd. 223, pp. 201-206, 1993.

36. S. Vyazovkin und A. Lesnikovich, „An approach to the solution of the inverse kinetic problem in the case of complex processes Part 1. Methods employing a series of thermalytical curves," Thermochimica Acta, Bd. 165, pp. 273-280, 1990.

37. T. Ozawa, „Kinetic analysis of derivative curves in thermal analysis," Journal of Thermal Analysis, Bd. 2, pp. 301-324, 1970.

38. K. Nawa, S. Ueda und H. Watanabe, „Advanced COPNA-resin as a low temperature curing resin for high-density electronic packages," IEEE Transactions on Components, Packaging, and Manufacturing Technology, Part B: Advanced Packaging, Bd. 19, Nr. 3, pp. 585-592, 1996.

39. R. A. Kirchhoff, C. J. Carriere, K. J. Bruza, N. G. Rondan und R. L. Sammler, „Benzocyclobutenes: a new class of high performance polymers," J. Macromol. Sci.- chem., Bd. A28, pp. 1079-1113, 1991.

40. S. Hogmark, S. Jacobson und M. Larsson, „Design and evaluation of tribological coatings," Wear, Bd. 246, pp. 20-33, 2000.

41. D. Dollimore, „The application of thermal analysis in studying the thermal decomposition of solids," Therrnochimica Acta, Bd. 203, pp. 7-23, 1992.

42. M. Wagner, G. Czerlinskil und M. Pring, „The application of fitting methods to chemical kinetics," Compur. Biol Med. Pergamon Press 1975. Vol 5. pp 105-117., Bd. 5, pp. 105-117, 1975.

43. A. A. Skordos und I. K. Pazitridge, „Cure Kinetics Modeling of Epoxy Resins Using a Non-Pa rametric Nu me rica I Procedure," Polymer Engineering and Science, Bd. 41, Nr. 5, pp. 793-805, May 2001.

44. S. Vyazovkin und C. A. Wight, „Model-free and model-fitting approaches to kinetic analysis of isothermal and nonisothermal data," Thermochimica Acta, Bd. 340, pp. 53-68, 1999.

13

COMPLEX MICROSTRUCTURES OF ABC TRIBLOCK COPOLYMER THIN FILMS DIRECTED BY POLYMER BRUSHES BASED ON SELF-CONSISTENT FIELD THEORY

Zhibin Jiang, Chang Xu, Yu dong Qiu, Xiaoliang Wang, Dongshan Zhou and Gi Xue

Department of Polymer Science and Engineering, Key Laboratory of High Performance Polymer Materials and Technology of Ministry of Education, State Key Laboratory of Coordination Chemistry, Nanjing National Laboratory of Microstructures, School of Chemistry and Chemical Engineering, Nanjing University, Nanjing 210093, China

ABSTRACT

The morphology and the phase diagram of ABC triblock copolymer thin film directed by polymer brushes are investigated by the self-consistent field theory in three dimensions. The polymer brushes coated on the substrate can be used as a good soft template to tailor the morphology of the block copolymer thin films compared with those on the hard substrates. The polymer brush is identical with the middle block B. By continuously changing the composition of the block copolymer, the phase diagrams are constructed for three cases with the fixed film thickness and the brush density: identical interaction parameters, frustrated and non-frustrated cases. Some ordered complex morphologies are observed: parallel lamellar phase with hexagonally packed pores at surfaces ($LAM_3^{//}$-HFs), perpendicular lamellar phase with cylinders at the interface ($LAM^{\perp}$-CI), and perpendicular hexagonally packed cylinders phase with rings at the interface ($C_2^{\perp}$-RI). A desired direction (perpendicular or parallel to the coated surfaces) of lamellar phases or cylindrical phases can be obtained by varying the composition and the interactions between different blocks. The phase diagram of ABC triblock copolymer thin film wetted between the polymer brush-coated surfaces is very useful in designing the directed pattern of ABC triblock copolymer thin film.

KEYWORDS

ABC triblock copolymer, Polymer brush, Morphology, Thin film, Self-consistent field theory, Phase diagram

1. BACKGROUND

Block copolymers consisting of chemically distinct polymers linked by a covalent bond at one end have the ability to self-assemble into a variety of ordered nanostructures such as lamellae (LAM), hexagonally packed cylinders (HEX), and body-centered cubic (BCC) spheres and more complex structures such as gyroid (G) in melts and solutions[1–7]. This unique characteristic of block copolymers provides possibilities for their potential applications in nanoscience, such as molecular template and nanotubes. Therefore, block copolymers have attracted a great deal of attention both in theory and experiment.

Self-assembly and phase separation in diblock copolymers have been well studied both theoretically and experimentally in the last few decades[8–14]. The phase behavior of diblock copolymers melts confined in a parallel slit or in a thin film has been extensively studied[15–25]. When the number of distinct blocks increases from two, i.e., ABC triblock copolymer, the complexity and variety of self-assembled structures are increased dramatically[1, 26–39]. If a surface or interface exists, the microdomain morphologies and the kinetics of microdomain ordering can change significantly. The complex and rich phase behaviors depend not only on molecular parameters, such as the interaction energies between distinct blocks and the architectures of block copolymers, but also on external variables, such as electric fields[40, 41], chemically patterned substrates[42–50], and interfacial interactions[4, 51–54]. The ABC linear triblock copolymer thin films confined between two hard walls have been intensively investigated theoretically[55–58]. Feng and Ruckenstein[59] studied ABC melts in thin films by Monte Carlo simulations and showed that the microdomain morphology can be very complicated and is affected by the composition, the interactions, and even the geometry of the confinement. Ludwigs et al.[60] observed a highly ordered hexagonally perforated lamella structure based on an ABC triblock copolymer thin film.

The previous work mainly concentrated on phases of several compositions of ABC triblock copolymer by varying the film thickness or the interfacial interaction. As we know, the polymer brush-coated surface is good from the energy view[30, 31]. It is equivalent to changing the surface-polymer interaction as polymer brush acts as a soft surface[30, 31, 61, 62]. Experimentally, random copolymers were used to control the wetting behavior of block copolymer[63, 64]. The results showed that the ordered structures can be easily obtained by changing the property of the surfaces or substrate, i.e., the interaction between the polymer and the surfaces. Ren et al.[61, 62] observed the structure transformation of the AB diblock copolymer thin film by tailoring the grafting density of the coated surface or the concentration of the copolymer. In order to know the whole phase behavior of ABC triblock copolymer thin film confined between two parallel polymer brush-coated surfaces, we use a combinatorial screening method based on the real space implementation of the self-consistent field theory (SCFT), originally proposed by Drolet and Fredrickson for block copolymer melts[65, 66, 57, 58] to search the equilibrium microphases of ABC linear triblock copolymers confined between the two parallel polymer brush-coated hard surfaces in three dimensions. In the present work, we concentrate on the thin film regime with film thickness of several R_{g0}. By continuously varying the compositions of the block copolymer, the morphologies are obtained, and the phase diagrams are constructed for three different cases of interaction parameters: (1)

identical interactions between three different components, (2) frustrated condition, and (3) non-frustrated condition.

2. METHODS

We assume the ABC triblock copolymer melt is confined between two parallel polymer brush-coated hard surfaces with a distance L_z along the z-axis. There are n_c ABC triblock copolymers with polymerization degree N and n_g polymer with polymerization degree P (here, we take $P = N$) grafting on the two parallel surfaces. Each copolymer chain consists of Nsegments with compositions (average volume fractions) f_A and f_B ($f_C = 1 - f_A - f_B$), respectively. The ABC triblock copolymer and the grafted polymers (brush) are assumed to be flexible, and the mixture is incompressible with each polymer segment having a statistical length a and occupying a fixed volumeρ−10ρ_0^{-1}. The two parallel surfaces coated by the polymer brush are horizontally placed on the xy-plane at $z = 0$ and $L_z + a$, respectively. The volume of the system is $V = L_x L_y L_z$, where L_x and L_y are the lateral lengths of the surfaces along the xy-plane and L_z is the film thickness. The grafting density is defined as $\sigma = n_g a^2/(2L_x L_y)$. The average volume fractions of the grafted chains and copolymers are $\varphi_g = n_g N/\rho_0 V$ and $\varphi_c = n_c N/\rho_0 V$, respectively.

In the SCFT, one considers the statistics of a single copolymer chain in a set of effective external fields w_i, where i represents block species A, B, and C or grafted polymers. These external fields, which represent the actual interactions between different components, are conjugated to the segment density fields, ϕ_i, of different species i. Hence, the free energy (in unit of $k_B T$) of the system is given by

$$F = -\varphi_c \ln(Q_c/\varphi_c V) - \varphi_g \ln\left(Q_g/\varphi_g V\right) - 1/V \int d\mathbf{r}\left[\sum_i w_i\phi_i + \xi\left(1 - \sum_i \phi_i\right)\right] + 1/V \int d\mathbf{r}\left[\frac{1}{2}\sum_{i \neq j} \chi_{ij} N \phi_i \phi_j + \sum_i H_{iS} N \phi_i \delta_{\mathbf{r},\mathbf{r}_s}\right] \quad (1)$$

where χ_{ij} is the Flory-Huggins interaction parameter between species i and j, ξ is the Lagrange multiplier (as a pressure), η_{iS} is the interaction parameter between the species i and the hard surface S. r_S is the position of the hard surfaces. $Q_c = \int dr q_c(r, 1)$ is the partition function of a single copolymer chain in the effective external fields w_A, w_B, and w_C, and $Q_g = \int dr q_g(r, 1)$ is the partition function of a grafted polymer chain in the external field w_g. The fundamental quantity to be calculated in mean-field studies is the polymer segment probability distribution function, $q(r, s)$, representing the probability of finding segment s at position r. It satisfies a modified diffusion equation using a flexible Gaussian chain model

$$\frac{\partial}{\partial s}q(\mathbf{r},s) = \frac{Na^2}{6}\nabla^2 q(\mathbf{r},s) - w(\mathbf{r})q(\mathbf{r},s) \tag{2}$$

where $w(r)$ is $w_A(r)$ when $0 < s < f_A$, $w_B(r)$ when $f_A < s < f_A + f_B$, $w_C(r)$ when $f_A + f_B < s < 1$ for ABC triblock copolymer, and $w_g(r)$ for the grafted polymer. The initial condition of Equation (2) satisfies $q_c(r, 0) = 1$ for ABC triblock copolymer. Because the two ends of the block copolymer are different, a second distribution function $q_c^+(\mathbf{r}, s)$ is needed which satisfies Equation (2) but with the right-hand side multiplied by -1 and the initial condition $q_c^+(\mathbf{r}, 1) = 1$. The initial condition of $q_g(r, s)$ for grafted polymer is $q_g(r, 0) = \delta(r - r_S)$, where r_S represents the position of the substrates, and that of $q_g^+(\mathbf{r}, s)$ is $q_g^+(\mathbf{r}, 1) = 1$. The periodic boundary condition is used for $q_c(\mathbf{r}, s)$, $q_c^+(\mathbf{r}, s)$, $q_g(\mathbf{r}, s)$, and $q_g^+(\mathbf{r}, s)$ along x- and y-directions when $z \in [0, L_z]$. $q_c(\mathbf{r}, s)$, $q_c^+(\mathbf{r}, s)$, $q_g(\mathbf{r}, s)$, and $q_g^+(\mathbf{r}, s)$ are equal to zero when $z \leq 0$ or $z \geq L_z$.

Minimization of the free energy with respect to density, pressure, and fields, $\delta F/\delta\phi = \delta F/\delta\xi = \delta F/\delta w = 0$, leads to the following equations.

$$w_A(\mathbf{r}) = \sum_{i \neq A} \chi_{Ai} N \phi_i(\mathbf{r}) + \xi(\mathbf{r}) + H_{AS} N \delta_{\mathbf{r},\mathbf{r}_s} \tag{3}$$

$$w_B(\mathbf{r}) = \sum_{i \neq B} \chi_{Bi} N \phi_i(\mathbf{r}) + \xi(\mathbf{r}) + H_{BS} N \delta_{\mathbf{r},\mathbf{r}_s} \tag{4}$$

$$w_C(\mathbf{r}) = \sum_{i \neq C} \chi_{Ci} N \phi_i(\mathbf{r}) + \xi(\mathbf{r}) + H_{CS} N \delta_{\mathbf{r},\mathbf{r}_s} \tag{5}$$

$$w_{\rm g}(\mathbf{r}) = \sum_{i \neq \rm g} \chi_{{\rm g}i} N \phi_i(\mathbf{r}) + \xi(\mathbf{r}) + \chi_{\rm gS} N \delta_{\mathbf{r},\mathbf{r}_s} \tag{6}$$

$$\phi_{\rm A}(\mathbf{r}) + \phi_{\rm B}(\mathbf{r}) + \phi_{\rm C}(\mathbf{r}) + \phi_{\rm g}(\mathbf{r}) = 1 \tag{7}$$

$$\phi_{\rm A}(\mathbf{r}) = \frac{\varphi_{\rm c} V}{Q_{\rm c}} \int_0^{f_{\rm A}} {\rm d}s q_{\rm c}(\mathbf{r}, s) q_{\rm c}^{+}(\mathbf{r}, s) \tag{8}$$

$$\phi_{\rm B}(\mathbf{r}) = \frac{\varphi_{\rm c} V}{Q_{\rm c}} \int_{f_{\rm A}}^{f_{\rm A}+f_{\rm B}} {\rm d}s q_{\rm c}(\mathbf{r}, s) q_{\rm c}^{+}(\mathbf{r}, s) \tag{9}$$

$$\phi_{\rm C}(\mathbf{r}) = \frac{\varphi_{\rm c} V}{Q_{\rm c}} \int_{f_{\rm A}+f_{\rm B}}^{1} {\rm d}s q_{\rm c}(\mathbf{r}, s) q_{\rm c}^{+}(\mathbf{r}, s) \tag{10}$$

$$\phi_{\rm g}(\mathbf{r}) = \frac{\varphi_{\rm g} V}{Q_{\rm g}} \int_0^1 {\rm d}s q_{\rm g}(\mathbf{r}, s) q_{\rm g}^{+}(\mathbf{r}, s) \tag{11}$$

Equations (3) to (11) form a close set of self-consistent equations, which are numerically implemented by a combinatorial screening algorithm proposed by Drolet and Fredrickson[65, 66]. The algori3thm consists of randomly generating the initial values of the fields w_i (r). Then, the diffusion equations are then integrated to obtain q and q^+, for $0 < s < 1$. The right-hand sides of Equations (8) to (11) are evaluated to obtain new values for the volume fractions of blocks A, B, and C, and grafted polymers. Moreover, the brief introduction of SCFT method can be found in some textbook, such as *Statistical Physics of Polymers: an Introduction* [67].

The polymerization of ABC triblock copolymer is $N = 60$ and that of the grafted chains is the same with the copolymers, i.e., $P = N = 60$. The grafting density of the grafted chains is set as $\sigma = 0.15$ and 0.2 to insure that the polymer brush is in the dry brush regime ($\sigma N^{1/2} > 1$)[68]. The interaction parameters $H_{i\,\rm S}$ (i = A, B, C) between the surfaces and the blocks are set to zero (the effect of the surface on the thin film is weakened because the surface is coated by polymer brushes), that means that the substrates are neutral. We only address the thin films of ABC triblock copolymer confined between densely polymer-grafted surfaces, and the grafted polymers are assumed to be identical with the middle block B. We continuously vary the compositions to search the morphology of the ABC block copolymer thin film. The simulations are performed on a 3D cubic box $L_x \times L_y \times L_z$. The two parallel hard surfaces are presented as planes at $z = 0$ and $L_z + a$, and the film thickness is set to $L_z = 40a$, which is appropriate for thin film with the effective thickness of several R_g. L_x and L_y along *xy*-plane can be varied between 40 to 45a to avoid the size effect and obtain the stable and perfect morphology. It should be noted that the resulting microphases largely depend on the initial conditions. Therefore, all the simulations are repeated many times using different random states to guarantee the structure is not occasionally observed. In this work, three cases are considered: (1) identical interactions between three different components, $\chi_{AB}N = \chi_{BC}N = \chi_{AC}N = 35$, which are widely studied in many theoretical works; (2) frustrated condition $\chi_{AB}N = \chi_{BC}N = 35$ and $\chi_{AC}N = 13$; and (3) non-frustrated condition, $\chi_{AB}N = \chi_{BC}N = 13$ and $\chi_{AC}N = 35$ based on the work of Jung[69] and Tyler[1]. Furthermore, the effect of the brush density is also included in the case of $\chi_{AB}N = \chi_{BC}N = \chi_{AC}N = 35$, which is actually equivalent to changing the effective film thickness.

3. RESULTS AND DISCUSSION

Figure 1 presents the morphologies of the ABC triblock copolymer thin film by varying the compositions of the block copolymer. The microphase patterns, displayed in the form of density, are the red, green, and blue, assigned to blocks A, B, and C, respectively. The final color plotted at each point is the mixture of three colors, in which the concentration of each color is proportional to the local volume fraction of an individual block. The 3D morphology can only give the three faces (*xy, yz, xz*) of the ABC triblock copolymer thin film. For some morphologies, the 3D isosurface graphs are also given for a clear view. The red, green, and blue colors in isosurface graphs are assigned to blocks A, B, and C for a good correspondence, respectively. In these 3D isosurface graphs, some only give one or two components. Here, we do not show the morphologies of the polymer brushes in order to clearly see the morphologies of the block copolymer. There are at least 15 stable morphologies found: two-color parallel lamellar phase ($LAM_2{}^{//}$), two-color perpendicular lamellar phase ($LAM_2{}^{\perp}$), three-color parallel lamellar phase ($LAM_3{}^{//}$), three-color perpendicular lamellar phase ($LAM_3{}^{\perp}$), parallel lamellar phase with hexagonally packed pores at surfaces ($LAM_3{}^{//}$-HFs), two-color parallel cylindrical phase ($C_2{}^{//}$), core-shell hexagonally packed spherical phase (CSHS), core-shell parallel cylindrical phase ($CSC_3{}^{//}$), perpendicular lamellar phase with cylinders at the interface ($LAM^{\perp}$-CI), perpendicular hexagonally packed cylinders phase with rings at the interface ($C_2{}^{\perp}$-RI), parallel lamellar phase with tetragonal pores ($LAM_3{}^{//}$-TF), perpendicular hexagonally packed cylindrical phase ($C_2{}^{\perp}$), sphere-cylinder transition phase (S-C), hexagonal pores (HF), and irregular lamellar phase (LAM_i). In these morphologies, there are some interesting structures, such as $LAM_3{}^{//}$-HFs, $LAM^{\perp}$-CI, $LAM_3{}^{//}$-TF, and HF. HF phase is also experimentally observed[60], which is very useful; for example, the perforated lamella can serve as a lithographic mask. There are two irregular phases, sphere-cylinder transition phase (S-C) and irregular lamellar phase (LAM_i). Due to the composition and the surface interaction competition, it is difficult to form the regular and stable phase. In fact, the parallel lamellar phases have three different arrangement styles near the brush. Because the brushes are identical to the middle block B, the block B should be near the brushes. But it is not always the case due to entropic effect. So, the blocks A, B, or C can be adjacent to the brushes. So in the following phase diagrams, we discern the three different arrangement styles of the parallel lamellar phases. When the block B is major in the block copolymer, the parallel lamellar phase with block B adjacent to brush layer is stable. When the block B is minor, the parallel lamellar phase with block A or B adjacent to brush layer is stable.

(1) Identical interaction parameter $\chi_{AB}N = \chi_{BC}N = \chi_{AC}N = 35$.

a. Influence of the composition

In this part, we consider the case of $\chi_{AB}N = \chi_{BC}N = \chi_{AC}N = 35$. Figure 2 gives the phase diagram of the ABC triblock copolymer when the brush density σ is 0.2. There are nine phases in the diagram. Due to the confinement of the ABC triblock copolymer and the tailoring effect of polymer brushes, the diagram is largely different from that in the bulk[33].

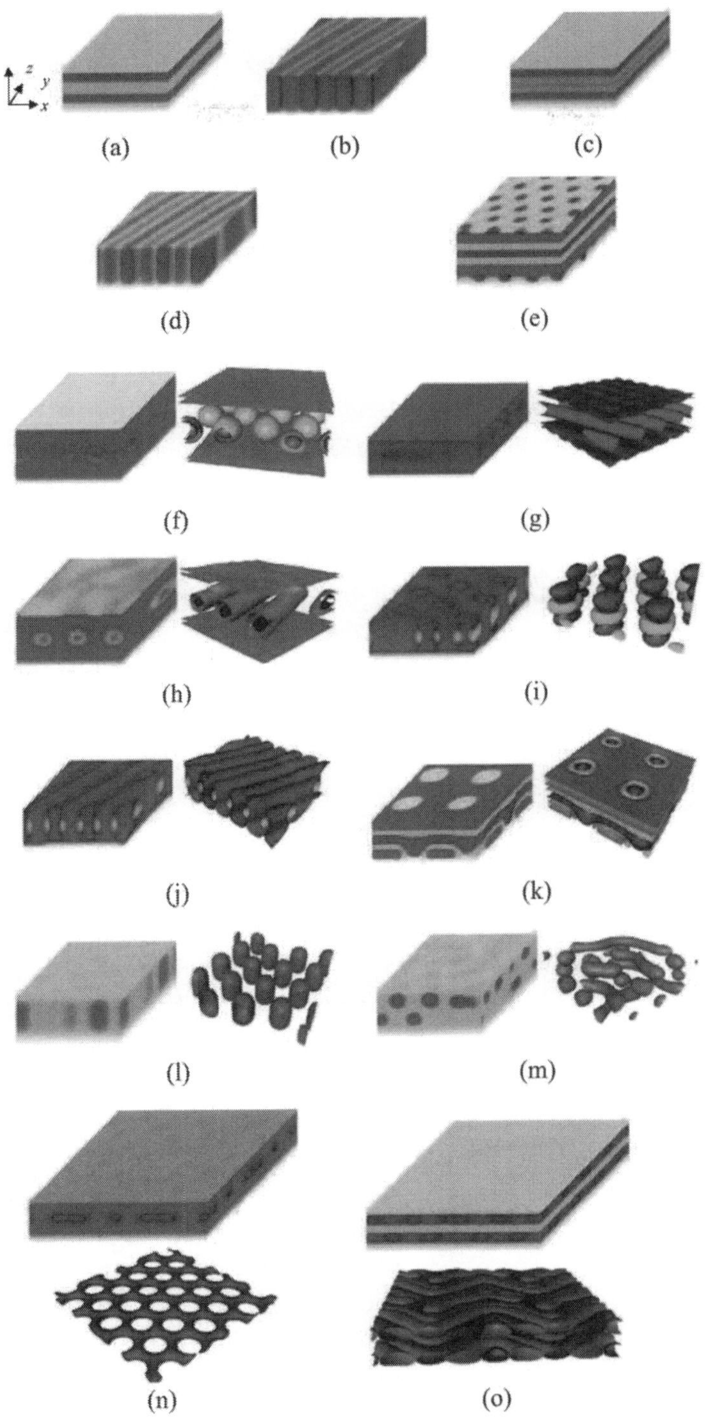
z
y
x
(a)
(b)
(c)
(d)
(e)
(f)
(g)
(h)
(i)
(j)
(k)
(l)
(m)
(n)
(o)

Figure 1 Morphologies of the ABC block copolymer thin film with $L_z = 40\ a$. The microphase patterns, displayed in the form of density, are the red, green, and blue, assigned to A, B, and C, respectively. For some morphologies, the 3D isosurface graphs are also given for a clear view beside the morphologies. The red, green, and blue colors in isosurface graphs are assigned to blocks A, B, and C for a good correspondence, respectively. **(a)** Two-color parallel lamellar phase ($LAM_2^{//}$), **(b)** two-color perpendicular lamellar phase ($LAM_2^{\perp}$), **(c)** three-color parallel lamellar phase ($LAM_3^{//}$), **(d)** three-color perpendicular lamellar phase ($LAM_3^{\perp}$), **(e)** parallel lamellar phase with hexagonally packed pores at surfaces ($LAM_3^{//}$-HFs), **(f)** core-shell hexagonally packed spherical phase (CSHS), **(g)** two-color parallel cylindrical phase ($C_2^{//}$), **(h)** core-shell parallel cylindrical phase ($CSC_3^{//}$), **(i)** perpendicular hexagonally packed cylindrical phase with rings at the interface ($C_2^{\perp}$-RI), **(j)** perpendicular lamellar phase with cylinders at the interface ($LAM^{\perp}$-CI), **(k)** parallel lamellar phase with tetragonal pores ($LAM_3^{//}$-TF), **(l)** perpendicular hexagonally packed cylindrical phase ($C_2^{\perp}$), **(m)** sphere-cylinder transition phase (S-C), **(n)** hexagonal pores (HF), and **(o)** irregular lamellar phase (LAM_i). Morphologies in **(n)** and **(o)** are enlarged by two times along *x*- and *y*-directions.

The disordered phase (Dis) exists at the three corners of the phase diagram. When the volume fractions of the three blocks are comparable, the three-color lamellar phase forms, which is similar with that in the bulk[33]. When one of the blocks is the minority, the phase behavior is similar with that of the diblock copolymer. When the middle block B is the minority, most of block B will accumulate between the A/C interface, and while the end block A or C is the minority, other block C or A will distribute in the middle block B to form one phase. There are many two-color phases near the edges of the phase diagram, and at this time, the lamellar phase is parallel to the surfaces. This shows that we can add a small functional block A or C to symmetric BC or AB diblock copolymer to obtain a lamellar phase parallel to the surfaces.

The diagram has the A-C reflectivity due to the symmetric architecture and the symmetric interaction parameters. When the volume fractions of the blocks A and C are comparable, the perpendicular lamellar phase easily forms except $f_B \leq 0.1$ and $f_B \geq 0.7$ and the compositions $f_A = 0.3$, $f_B = 0.3, f_C = 0.4$, and $f_A = 0.4, f_B = 0.3, f_C = 0.3$.

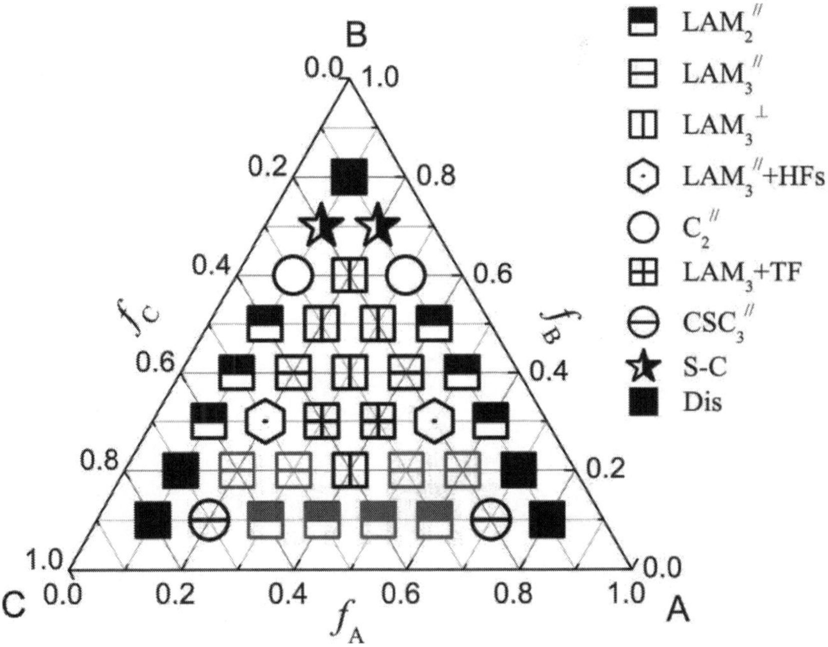

Figure 2 Phase diagram of ABC triblock copolymer with $\chi_{AB}N = \chi_{BC}N = \chi_{AC}N = 35$ at grafting density $\sigma = 0.20$. Dis represents the disordered phase. The red, blue, or black icons showing the parallel lamellar phases discern the different arrangement styles of the block copolymer with block A, block C, or block B adjacent to the brush layers, respectively.

b. Influence of the grafting density

We also consider the grafting density $\sigma = 0.15$ when $\chi_{AB}N = \chi_{BC}N = \chi_{AC}N = 35$. The grafting density decreases a little, which shows that the effective film thickness increases. The phase diagram is shown in Figure 3. From the figure, we can see that the lamellar phase region contracts and some new phases emerge, such as two-color perpendicular lamellar phase ($LAM_2^{\perp}$) and core-shell hexagonally packed spherical phase (CSHS). Due to the decrease of the grafting density, the influence of the brush will weaken. Similar with the case of $\sigma = 0.20$, the core-shell structures occur near the corners A and C. CSHS phase forms at $f_A = 0.10$, $f_B = 0.10$, $f_C = 0.80$; $f_A = 0.80$, $f_B = 0.10$, $f_C = 0.10$. The core-shell cylindrical phase occurs near the phase CSHS. In these cases, the block A (or C) forms the majority, the block C (or A) forms the 'core,' and the middle block B is around the block C (or A) forming the 'shell' of the core.

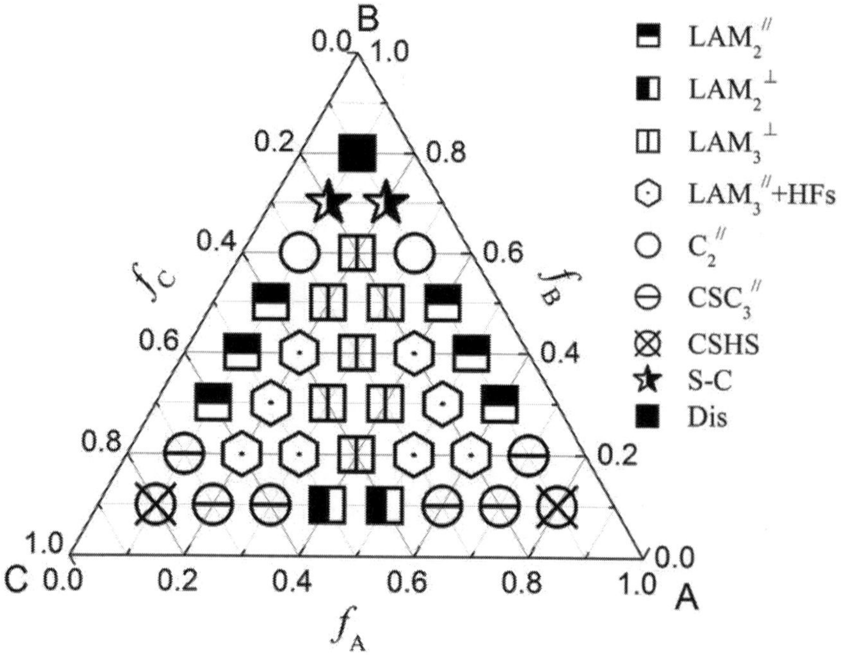

Figure 3 Phase diagram of ABC triblock copolymer with $\chi_{AB}N = \chi_{BC}N = \chi_{AC}N = 35$ at grafting density $\sigma = 0.15$. Dis represents the disordered phase.

Comparing the phase diagram with that in the bulk[33], the direction of the lamellar phase can be tailored by changing the grafting density when the middle blocks are the minority and the ABC triblock copolymer is symmetric, i.e. $f_A = f_C$. The parallel lamellar phase with hexagonally packed pores at surfaces ($LAM_3{}^{//}$-HFs) can easily form at some compositions.

In general, the block copolymer experiences the film confinement under this condition. Moreover, the block copolymer experiences the brush polymer tailoring, especially at the interface between the block copolymer and the polymer brush. Therefore, some new phases form, and the phase diagram is more complicated. Even for the lamellar phase, there are two styles: the perpendicular and parallel ones. The perpendicular lamellar phase always occurs when the volume fractions of the three components are comparable. The parallel lamellar phase forms at the middle edge of the phase diagram in most cases. From the above two phase diagrams, we can see that the hexagonally packed pores at the interface between the block copolymers and the polymer brush-coated surfaces occur. It is very useful in designing thin films with functional dots.

(2) Frustrated case $\chi_{AB}N = \chi_{BC}N = 35, \chi_{AC}N = 13$

It is energetically unfavorable when $\chi_{AC}N << \chi_{AB}N \approx \chi_{BC}N$; that is to say, the repulsive interaction between the two ends is the smallest in the three interaction parameters. Thus, the block B has to be limited in spheres, rings, or cylinders to increase the contacting interface between the blocks A and C. Some phases are also observed in the frustrated ABC triblock copolymer in bulk[70], such as $C_2{}^{\perp}$-RI and $LAM^{\perp}$-CI (which correspond to C + HEL and L + C(II) in[56]). The phase diagram is shown in Figure 4 for $\chi_{AB}N = \chi_{BC}N = 35$ and $\chi_{AC}N = 13$.

Due to the energetic confinement, the two-color lamellar phase is easy to form. When the middle block B is the minority, the phases are complex. The block B will accumulate near the interface between the blocks A and C, which can be comparable with that in the bulk in the frustrated case[33, 70]. For the symmetric ABC triblock copolymer, i.e., $f_A = f_C$, with the increase of the volume fraction of the middle block B, the phase will change from the perpendicular lamellar phase to perpendicular lamellar phase with cylinders at the interface to irregular lamellar phase to three-color parallel lamellar phase. This shows that the direction of the lamellar phase can be tailored. The irregular lamellar phase (three points $f_A = 0.3$, $f_B = 0.3$, $f_C = 0.4$; $f_A = 0.4$, $f_B = 0.3$,

$f_C = 0.3$; $f_A = 0.3, f_B = 0.4, f_C = 0.3$) forms because of two reasons: one is the three blocks with almost equal volume fraction, and the middle block B will stay near to the polymer-coated (same with block B) substrates, so there is not enough block B to form the perfect lamellar phase. The other reason is $\chi_{AC}N << \chi_{AB}N \approx \chi_{BC}N$, then the copolymer chain will overcome the elastic energy to form the A/C interface. Therefore, the phase is not perfect because of the composition competition and the energy competition. And the most important is that perpendicular hexagonally packed cylindrical phase with rings at the interface ($C_2^{\perp}$-RI) and perpendicular lamellar phase with cylinders at the interface ($LAM^{\perp}$-CI) occur in this frustrated case, see Figure 1i,j. In fact, these two phases are obtained in the frustrated ABC triblock copolymer with interaction parameters $\chi_{AB}N = \chi_{BC}N = 35$ and $\chi_{AC}N = 15$ in bulk[70].

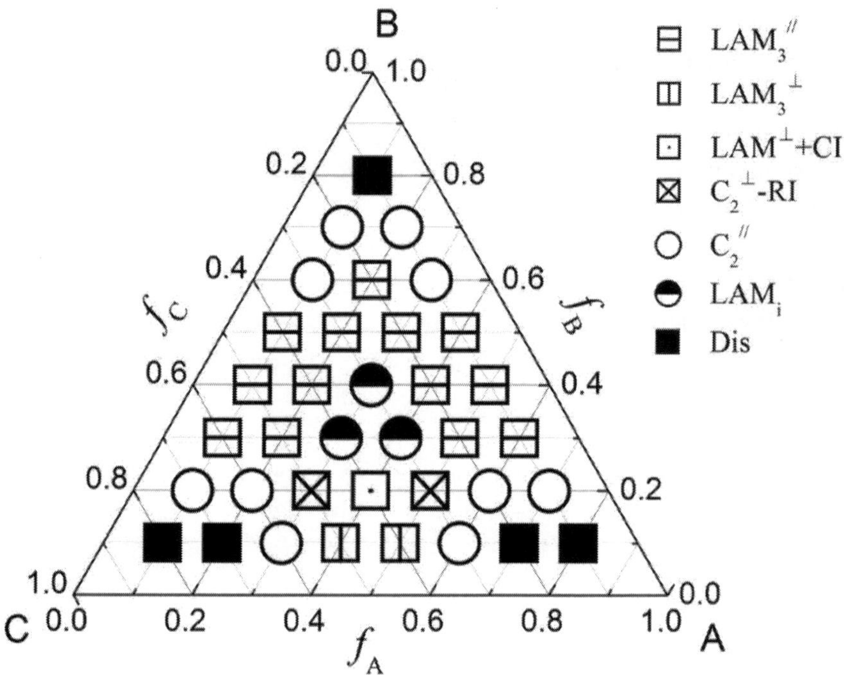

Figure 4 Phase diagram of ABC triblock copolymer with $\chi_{AB}N = \chi_{BC}N = 35$ and $\chi_{AC}N = 13$ at grafting density $\sigma = 0.2$. Dis represents the disordered phase.

(3) Non-frustrated case ($\chi_{AB} N = \chi_{BC} N = 13, \chi_{AC} N = 35$)

It is an energetically favorable case when the repulsive interaction between the end blocks A and C is larger than that for blocks A and B or blocks B and C. Here, we consider the case of $\chi_{AB}N = \chi_{BC}N = 13$ and $\chi_{AC}N = 35$, which is used when considering the non-frustrated case for ABC block copolymer[1]. The phase diagram of ABC triblock copolymer thin film for $\chi_{AB}N = \chi_{BC}N = 13$ and$\chi_{AC}N = 35$ is shown in Figure 5. Eight phases are found in this case. Due to the relative weak interaction between the blocks A and B and between the blocks B and C, the disordered phase occurs at the corners of the three blocks. The lamellar phase region is very large. The three-color lamellar phase forms when the volume fractions of the three components are comparable. The two-color lamellar phase is stable in the middle of the three edges in the phase diagram. When the volume fractions of the blocks A and C are equal, the perpendicular lamellar phase easily forms between $f_B = 0.1$ and 0.6.

(4) Comparison with ABC triblock copolymer thin film without polymer brush-coated substrates

In this part, we give two cases for comparison between the ABC triblock copolymer thin film with and without polymer brush-coated substrates ($\sigma = 0.15$) at $\chi_{AB}N = \chi_{BC}N = \chi_{AC}N = 35$. In order to simulate the similar interface environment with the ABC triblock copolymer thin film between polymer brush-coated substrates, we set the interaction parameters $\eta_{AS}N = \eta_{CS}N = 35$ and$\eta_{BS}N = 0$ for the ABC triblock copolymer thin film between hard surfaces, which means the substrate is good for the middle block B. In principle, the effective film thickness for the ABC triblock copolymer thin film confined between the polymer brush-coated substrates is like $L_z^{eff} = L_z - 2a\sigma P$ for $\sigma P^{1/2} > 1$ (where 2 is just for the upper and lower polymer grafted surfaces, brush height $h = a\sigma P$ for $\sigma P^{1/2} > 1$[68]). When the ABC triblock copolymer is confined between two hard surfaces (without polymer brush-coated substrates), the corresponding effective film thickness is $22a$ in this case.The morphology comparison of ABC triblock copolymer confined between polymer-coated substrates and hard surfaces is listed in Figure 6. The first column is the composition of ABC triblock copolymer. The second column is the morphologies of the ABC triblock copolymer confined between the polymer brush-coated surfaces and the morphologies of the polymer brush. The third column is the morphologies of ABC triblock copolymer confined between hard surfaces (without polymer brush-coated) and the 3D isosurface for a clear view. The microphase patterns, displayed in the form of density, are the red, green, and blue, assigned to A, B, and C, respectively. Similarly, the red, green, and blue colors in 3D isosurface graphs are assigned to blocks A, B, and C for a good correspondence, respectively. For the ABC triblock copolymer confined between polymer brush-coated substrates, the morphology of the grafted polymer on the lower substrate (polymer brush) is also shown below the morphology of ABC triblock copolymer. We only give the morphology of the grafted polymer on the lower substrate (polymer brush) due to the symmetry of the polymer brush (the two polymer brush-coated surfaces are identical). For the ABC triblock copolymer confined between the hard surfaces, the 3D isosurface is also shown below the morphology.

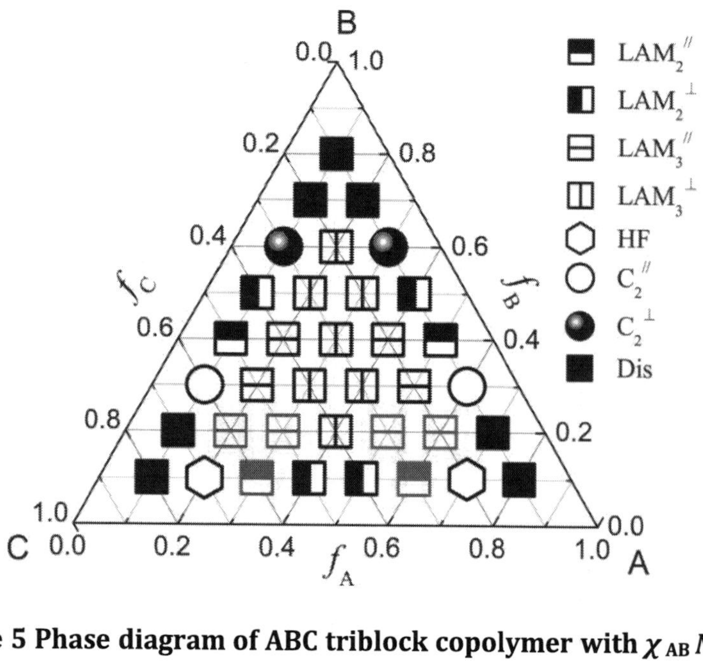

Figure 5 Phase diagram of ABC triblock copolymer with $\chi_{AB}N = \chi_{BC}N = 13$ and $\chi_{AC}N = 35$ at grafting density $\sigma = 0.2$. Dis represents the disordered phase. The red, blue, or black icons showing the parallel lamellar phases discern the different arrangement styles of the block copolymer with block **A**, block **C**, or block **B** adjacent to the brush layers, respectively.

Composition f_A, f_B, f_C	Morphology of ABC triblock copolymer confined between polymer-coated substrates	Morphology of ABC triblock copolymer confined between hard surfaces
0.4,0.4,0.2	 ABC 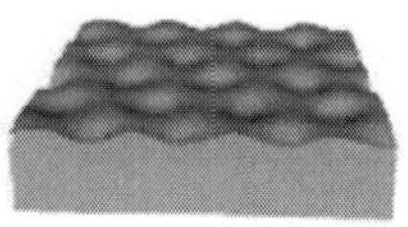 Polymer brush	 ABC  3D isosurface
0.4,0.2,0.4	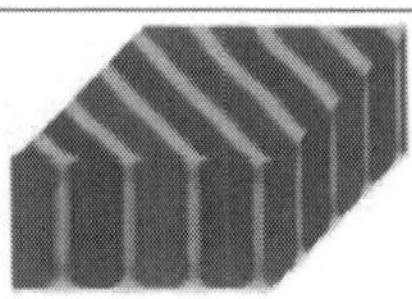 ABC  Polymer brush	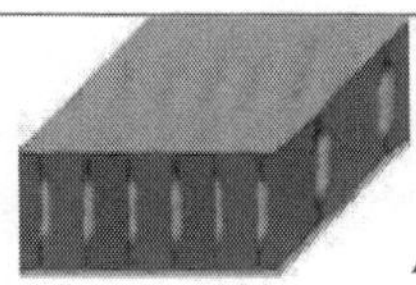 ABC  3D isosurface

Figure 6 Comparison of the morphology of ABC triblock copolymer confined between hard surfaces and polymer brush-coated substrates. The microphase patterns, displayed in the form of density, are the red, green, and blue, assigned to A, B, and C, respectively. The 3D isosurface graphs are also given for a clear view for the ABC triblock copolymer confined between the hard surfaces. The red, green, and blue colors in isosurface graphs are assigned to the blocks A, B, and C for a good correspondence, respectively. For the ABC triblock copolymer confined between polymer brush-coated substrates, the 3D isosurface of the grafted polymer on the lower substrate is also shown below the morphology due to the symmetry of the polymer brush. For the ABC triblock copolymer confined between hard surfaces, the 3D isosurface is also shown below the morphology.

When $f_A = 0.4$, $f_B = 0.4$, and $f_C = 0.2$ at $\sigma = 0.15$, the phase $LAM_3^{//}$-HFs is stable, while the stable phase for the thin film confined between hard surfaces is three-color lamellae with parallel cylinders at the interfaces. When $f_A = 0.4$, $f_B = 0.2$, and $f_C = 0.4$ at $\sigma = 0.15$, the perpendicular lamellar phase $LAM_3^{\perp}$ is stable, while the perpendicular lamellar phase with cylinder at the interfaces is stable without the coated polymer brush at the surfaces. From the morphology of the polymer brush, we can see that there is some ordered pattern at the interface between the thin film and the polymer brush. So, we think the coated polymers on the substrates have penetrated into the ABC triblock copolymer thin film, and the interaction between them contributes to morphology formation of the thin film. For the case of $f_A = 0.4$, $f_B = 0.2$, and $f_C = 0.4$, the perpendicular lamellar phase with cylinders at the interfaces is stable without the coated polymer brush at the surfaces. But when it is confined between the polymer brush-coated substrates, the polymer brush will penetrate into the block copolymer thin film and form one phase with the middle block B, so the perpendicular lamellar phase occurs.

The density profile of the block copolymer along z-direction can be obtained by $\phi_i(z) = \sum_{x,y} \phi_i(x, y, z)/(L_x L_y/a^2)$ ($i$belongs to blocks A, B, and C and grafting polymer g). Figure 7 gives the density profiles of the blocks A(solid), B(dash), and C(dot) and the grafting polymer(dash dot) for the cases (a) $f_A = 0.4$, $f_B = 0.4$, and $f_C = 0.2$ and (b) $f_A = 0.4$, $f_B = 0.2$, $f_C = 0.4$. The polymer brush and the middle block B have interpenetration. So, the interfacial morphology is different from the block copolymer confined between hard surfaces. We can see the lamellar distribution parallel to the substrates for $f_A = 0.4$, $f_B = 0.4$, $f_C = 0.2$, so there are peaks along z-direction which correspond to the domain centers of the blocks. The perpendicular lamellar phase forms for $f_A = 0.4$, $f_B = 0.2$, $f_C = 0.4$, and the uniform distribution exists in the middle of the film. The curves for the blocks A and C are overlapped due to the same composition and the symmetric interaction parameters between different blocks. These density profiles clearly show the coated polymer brush attributes to the morphology formation of the block copolymer. We can use the polymer brush to tailor the morphology of the block copolymer thin film.

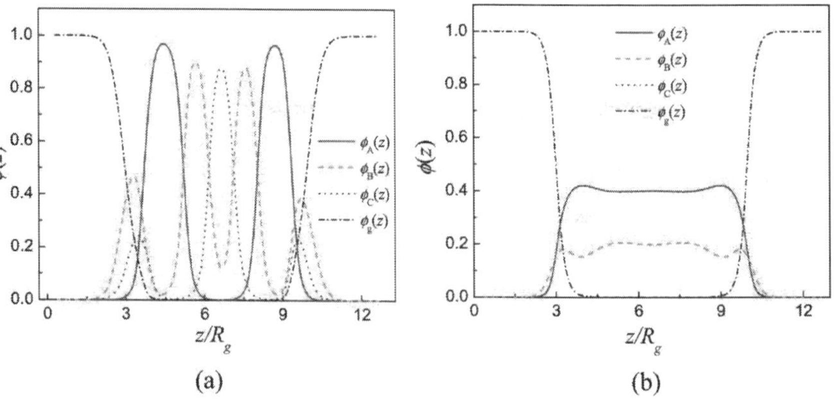

Figure 7 Density distribution of the different components along *z* -direction with $\chi_{AB}N=\chi_{BC}N=\chi_{AC}N=35$, $\sigma=0.15$. (a) $f_A=0.4, f_B=0.4, f_C=0.2$; **(b)** $f_A=0.4, f_B=0.2, f_C=0.4$.

4. CONCLUSIONS

The morphology and the phase diagrams of ABC triblock copolymer thin film confined between polymer brush-coated surfaces are investigated by the real-space self-consistent field theory in three dimensions. The coated polymer brush is identical with the middle block B. By continuously changing the composition of the block copolymer, the phase diagrams are constructed for three cases with the fixed film thickness $L_z=40a$ and the grafting density $\sigma=0.20$: (1) identical interactions between three different components, $\chi_{AB}N=\chi_{BC}N=\chi_{AC}N=35$; (2) frustrated condition $\chi_{AB}N=\chi_{BC}N=35$ and $\chi_{AC}N=13$; and (3) non-frustrated condition,$\chi_{AB}N=\chi_{BC}N=13$ and $\chi_{AC}N=35$. Furthermore, the brush density $\sigma=0.15$ is also included in the case of $\chi_{AB}N=\chi_{BC}N=\chi_{AC}N=35$. Fifteen stable morphologies are obtained: $LAM_2^{//}$, $LAM_2^{\perp}$, $LAM_3^{//}$, $LAM_3^{\perp}$, $LAM_3^{//}$-HFs, $C_2^{//}$, CSHS, $CSC_3^{//}$, $LAM^{\perp}$-CI, $C_2^{\perp}$-RI, $LAM_3^{//}$-TF, $C_2^{\perp}$, S-C, HF, and LAM_i. The morphology of the block copolymer thin film largely depends on the compositions and the surface interaction besides the film thickness. The complex morphology can be obtained at the energetically unfavorable condition, such as the cases for $\chi_{AB}N=\chi_{BC}N=\chi_{AC}N=35$ and $\chi_{AB}N=\chi_{BC}N=35$ and $\chi_{AC}N=13$. Although the grafted polymers are identical to the middle block B, the perpendicular lamellar phase is not always the stable one. The perpendicular or parallel lamellar phases can be obtained by varying the composition (besides changing the film thickness) and the interactions between different blocks. When one of the end block A or C is minority, the two-color parallel lamellar phase easily forms, while the perpendicular lamellar phase is stable when the block copolymer is symmetric, i.e., $f_A = f_C$. Even the direction of the cylinders can also be tuned for the non-frustrated case, where the direction of the cylinder can be tailored by the composition of the block. The parallel cylindrical phase forms if the end block A or C is the majority (f_A or $f_C = 0.6$), and the perpendicular cylindrical phase forms if the middle block B is the majority ($f_B = 0.6$) for the non-frustrated case. There are some interesting phases, such as hexagonally packed pores at surfaces ($LAM_3^{//}$ + HFs) and perpendicular hexagonally packed cylindrical phase with rings at the interface ($C_2^{\perp}$-RI). Compared with the case of the ABC triblock copolymer thin film without polymer brush-coated substrate, the morphologies of ABC triblock copolymer thin film confined between polymer brush-coated substrates show some preferences and are easily controllable. The perpendicular lamellar phase is stable when the block copolymer is symmetric, i.e., $f_A = f_C$ unlike the case of thin film without polymer brush-coated substrate, the direction will change at different film thickness. Although the grafted polymers are identical to the middle block B, the perpendicular lamellar phase is not always the stable one. The perpendicular or parallel lamellar phases can be obtained by varying the composition and the interactions between different blocks. Even the direction of the cylinders can also be tuned for the non-frustrated case. Our simulation results give an overview of ABC triblock copolymer thin film confined between the polymer brush-coated surfaces and are very useful in designing the complex morphology of ABC triblock copolymer thin film; for example, we can obtain the $LAM_3^{//}$ -HFs, which is potentially useful in designing the functional dots near the surfaces.

ACKNOWLEDGEMENTS

We gratefully acknowledge the financial support from the National Natural Science Foundation of China (Grant Nos. 20874046, 21074053, 21174062, and 51133002), National Basic Research Program of China (Grant no. 2010CB923303), the foundation research project of Jiangsu province (BK20131269) Fundamental Research Funds for the Central Universities (No.1095020515), and Program for Changjiang Scholars and Innovative Research Team in University.

COMPETING INTERESTS

The authors declare that they have no competing interests.

AUTHORS' CONTRIBUTIONS

ZBJ, CX, and YDQ carried out the simulations. ZBJ performed the data analysis and drafted the manuscript and participated in its design. XLW, DSZ, and GX participated in the design of the study and conceived of the study. All authors read and approved the final manuscript.

REFERENCES

1. Tyler CA, Qin J, Bates FS, Morse DC: SCFT study of nonfrustrated ABC triblock copolymer melts. *Macromolecules* 2007, 40: 4654–4668.

2. Hamley IW: *The Physics of Block Copolymer*. New York: Oxford University Press; 1998.

3. Hamley IW: Nanostructure fabrication using block copolymers. *Nanotechnology* 2003, 14: R39-R54.

4. Mansky P, Russell TP, Hawker CJ, Mays J, Cook DC, Satija SK: Interfacial segregation in disordered block copolymers: effect of tunable surface potentials. *Phys Rev Lett* 1997, 79: 237–240.

5. Liu X, Stamm M: Fabrication of highly ordered polymeric nanodot and nanowire arrays templated by supramolecular assembly block copolymer nanoporous thin films. *Nanoscale Res Lett* 2009, 4: 459–464.

6. Balsamo V, Collins S, Hamley IW: Nanopatterned surfaces obtained with semicrystalline ABC triblock copolymers. *Polymer* 2002, 43: 4207–4216.

7. Peponi L, Marcos-Fernandez A, Maria Kenny J: Nanostructured morphology of a random P(DLLA-co-CL) copolymer. *Nanoscale Res Lett*2012, 7: 1–7.

8. Srinivas G, Discher DE, Klein ML: Self-assembly and properties of diblock copolymers by coarse-grain molecular dynamics. *Nat Mater*2004, 3: 638–644.

9. Srinivas G, Shelley JC, Nielsen SO, Discher DE, Klein ML: Simulation of diblock copolymer self-assembly, using a coarse-grain model. *J Phys Chem B* 2004, 108: 8153–8160.

10. Glass R, Moller M, Spatz JP: Block copolymer micelle nanolithography. *Nanotechnology* 2003, 14: 1153–1160.

11. Sun RG, Wang YZ, Wang DK, Zheng QB, Kyllo EM, Gustafson TL, Wang FS, Epstein AJ: High PL quantum efficiency of poly(phenylene vinylene) systems through exciton confinement. *Synth Met* 2000, 111: 595–602.

12. Wurlitzer A, Politsch E, Huebner S, Kruger P, Weygand M, Kjaer K, Hommes P, Nuyken O, Cevc G, Losche M: Conformation of polymer brushes at aqueous surfaces determined with X-ray and neutron reflectometry. 2. High-density phase transition of lipopolyoxazolines.*Macromolecules* 2001, 34: 1334–1342.

13. Kumar R, Muthukumar M: Microphase separation in polyelectrolytic diblock copolymer melt: weak segregation limit. *J Chem Phys* 2007, 126: 214902.

14. Liu Z, Jiang ZB, Yang H, Bai SM, Wang R, Xue G: Crowding agent induced phase transition of amphiphilic diblock copolymer in solution.*Chin J Polym Sci* 2013, 31: 1491–1500.

15. Matsen MW: Electric field alignment in thin films of cylinder-forming diblock copolymer. *Macromolecules* 2006, 39: 5512–5520.

16. Morkved TL, Jaeger HM: Thickness-induced morphology changes in lamellar diblock copolymer ultrathin films. *Europhys Lett* 1997, 40: 643–648.

17. Geisinger T, Muller M, Binder K: Symmetric diblock copolymers in thin films. I. Phase stability in self-consistent field calculations and Monte Carlo simulations.*J Chem Phys* 1999, 111: 5241–5250.

18. Geisinger T, Muller M, Binder K: Symmetric diblock copolymers in thin films. II. Comparison of profiles between self-consistent field calculations and Monte Carlo simulations. *J Chem Phys* 1999, 111: 5251–5258.

19. Huinink HP, Brokken-Zijp JCM, van Dijk MA, Sevink GJA: Asymmetric block copolymers confined in a thin film. *J Chem Phys* 2000, 112: 2452–2462.

20. Sevink GJA, Zvelindovsky AV, Fraaije J, Huinink HP: Morphology of symmetric block copolymer in a cylindrical pore.*J Chem Phys* 2001, 115: 8226–8230.

21. Spontak RJ, Shankar R, Bowman MK, Krishnan AS, Hamersky MW, Samseth J, Bockstaller MR, Rasmussen KO: Selectivity- and size-induced segregation of molecular and nanoscale species in microphase-ordered triblock copolymers. *Nano Lett* 2006, 6: 2115–2120.

22. Turner MS: Equilibrium properties of a diblock copolymer lamellar phase confined between flat plates. *Phys Rev Lett* 1992, 69: 1788–1791.

23. Kellogg GJ, Walton DG, Mayes AM, Lambooy P, Russell TP, Gallagher PD, Satija SK: Observed surface energy effects in confined diblock copolymers. *Phys Rev Lett* 1996, 76: 2503–2506.

24. Lambooy P, Russell TP, Kellogg GJ, Mayes AM, Gallagher PD, Satija SK: Observed frustration in confined block-copolymers. *Phys Rev Lett*1994, 72: 2899–2902.

25. Walton DG, Kellogg GJ, Mayes AM, Lambooy P, Russell TP: A free-energy model for confined diblock copolymers. *Macromolecules* 1994, 27: 6225–6228.

26. Zhang XH, Berry BC, Yager KG, Kim S, Jones RL, Satija S, Pickel DL, Douglas JF, Karim A: Surface morphology diagram for cylinder-forming block copolymer thin films. *ACS Nano* 2008, 2: 2331–2341.

27. Feng J, Ruckenstein E: Self-assembling of ABC linear triblock copolymers in nanocylindrical tubes. *J Chem Phys* 2007, 126: 124902.

28. Xia Y, Chen J, Sun Z, Shi T, An L, Jia Y: Self-assembly of linear ABC coil-coil-rod triblock copolymers. *Polymer* 2010, 51: 3315–3319.

29. Park HW, Jung J, Chang T, Matsunaga K, Jinnai H: New epitaxial phase transition between DG and HEX in PS-b-PI. *J Am Chem Soc* 2009, 131: 46–47.

30. Wang R, Zhang S, Qiu Y: Hetero-structure of ABC triblock copolymer thin film on polymer-coated substrate. *Polymer* 2011, 52: 586–592.

31. Jiang ZB, Wang R, Xue G: Self assembly of ABC triblock copolymer thin films on a brush-coated substrate. *Chin J Polym Sci* 2009, 27: 583–592.

32. Matsen MW: Gyroid versus double-diamond in ABC triblock copolymer melts. *J Chem Phys* 1998, 108: 785–796.

33. Tang P, Qiu F, Zhang HD, Yang YL: Morphology and phase diagram of complex block copolymers: ABC linear triblock copolymers. *Phys Rev E* 2004, 69: 031803.

34. Dong Y-Q, Dong B-T, Du F-S, Meng J-Q, Li Z-C: An All ATRP route to PMMA-PEO-PS and PMAA-PEO-PS miktoarm ABC star terpolymer.*Polymer* 2009, 50: 125–132.

35. Lin B, Zhang H, Qiu F, Yang Y: Self-assembly of ABC star triblock copolymer thin films confined with a preferential surface: a self-consistent mean field theory. *Langmuir* 2010, 26: 19033–19044.

36. Kong W, Li B, Jin Q, Ding D, Shi A-C: Helical vesicles, segmented semivesicles, and noncircular bilayer sheets from solution-state self-assembly of ABC miktoarm star terpolymers. *J Am Chem Soc* 2009, 131: 8503–8512.

37. Yin Y, Jiang R, Li B, Jin Q, Ding D, Shi A-C: Self-assembly of grafted Y-shaped ABC triblock copolymers in solutions. *J Chem Phys* 2008, 129: 154903.

38. Tang P, Qiu F, Zhang HD, Yang YL: Morphology and phase diagram of complex block copolymers: ABC star triblock copolymers. *J Phys Chem B* 2004, 108: 8434–8438.

39. Wang C, Lee DH, Hexemer A, Kim MI, Zhao W, Hasegawa H, Ade H, Russell TP: Defining the nanostructured morphology of triblock copolymers using resonant soft X-ray scattering. *Nano Lett* 2011, 11: 3906–3911.

40. Morkved TL, Lu M, Urbas AM, Ehrichs EE, Jaeger HM, Mansky P, Russell TP: Local control of microdomain orientation in diblock copolymer thin films with electric fields. *Science* 1996, 273: 931–933.

41. Boker A, Muller AHE, Krausch G: Nanoscopic surface patterns from functional ABC triblock copolymers. *Macromolecules* 2001, 34: 7477–7488.

42. Ionov L, Minko S, Stamm M, Gohy JF, Jerome R, Scholl A: Reversible chemical patterning on stimuli-responsive polymer film: environment-responsive lithography. *J Am Chem Soc* 2003, 125: 8302–8306.

43. Tsori Y, Andelman D: Diblock copolymer ordering induced by patterned surfaces above the order–disorder transition. *Macromolecules*2001, 34: 2719–2727.

44. Li RR, Dapkus PD, Thompson ME, Jeong WG, Harrison C, Chaikin PM, Register RA, Adamson DH: Dense arrays of ordered GaAs nanostructures by selective area growth on substrates patterned by block copolymer lithography. *Appl Phys Lett* 2000, 76: 1689–1691.

45. Pereira GG, Williams DRM: Lamellar interface distortions of thin, commensurately aligned diblock copolymer films on patterned surfaces.*Langmuir* 1999, 15: 2125–2129.

46. Pereira GG, Williams DRM: Equilibrium properties of diblock copolymer thin films on a heterogeneous, striped surface. *Macromolecules*1998, 31: 5904–5915.

47. Pereira GG, Williams DRM: Diblock copolymer thin films on heterogeneous striped surfaces: commensurate, incommensurate and inverted lamellae. *Phys Rev Lett* 1998, 80: 2849–2852.

48. Ludwigs S, Schmidt K, Stafford CM, Amis EJ, Fasolka MJ, Karim A, Magerle R, Krausch G: Combinatorial mapping of the phase behavior of ABC triblock terpolymers in thin films: experiments. *Macromolecules* 2005, 38: 1850–1858.

49. Wolff M, Scholz U, Hock R, Magerl A, Leiner V, Zabel H: Crystallization of micelles at chemically terminated interfaces. *Phys Rev Lett* 2004, 92: 255501.

50. Park S, Lee DH, Xu J, Kim B, Hong SW, Jeong U, Xu T, Russell TP: Macroscopic 10-terabit-per-square- inch arrays from block copolymers with lateral order. *Science* 2009, 323: 1030–1033.

51. Luzinov I, Minko S, Tsukruk VV: Adaptive and responsive surfaces through controlled reorganization of interfacial polymer layers. *Prog Polym Sci* 2004, 29: 635–698.

52. Peters RD, Yang XM, Nealey PF: Morphology of thin films of diblock copolymers on surfaces micropatterned with regions of different interfacial energy. *Macromolecules* 2002, 35: 1822–1834.

53. Walton DG, Soo PP, Mayes AM, Allgor SJS, Fujii JT, Griffith LG, Ankner JF, Kaiser H, Johansson J, Smith GD, Barker JG, Satija SK: Creation of stable poly(ethylene oxide) surfaces on poly(methyl methacrylate) using blends of branched and linear polymers. *Macromolecules* 1997, 30: 6947–6956.

54. Ryu DY, Shin K, Drockenmuller E, Hawker CJ, Russell TP: A generalized approach to the modification of solid surfaces. *Science* 2005, 308: 236–239.

55. Pickett GT, Balazs AC: Equilibrium behavior of confined triblock copolymer films. *Macromol Theory Simul* 1998, 7: 249–255.

56. Chen HY, Fredrickson GH: Morphologies of ABC triblock copolymer thin films. *J Chem Phys* 2002, 116: 1137–1146.

57. Ludwigs S, Krausch G, Magerle R, Zvelindovsky AV, Sevink GJA: Phase behavior of ABC triblock terpolymers in thin films: mesoseale simulations. *Macromolecules* 2005, 38: 1859–1867.

58. Knoll A, Lyakhova KS, Horvat A, Krausch G, Sevink GJA, Zvelindovsky AV, Magerle R: Direct imaging and mesoscale modelling of phase transitions in a nanostructured fluid. *Nat Mater* 2004, 3: 886–890.

59. Feng J, Ruckenstein E: Monte Carlo simulation of triblock copolymer thin films. *Polymer* 2002, 43: 5775–5790.

60. Ludwigs S, Boker A, Voronov A, Rehse N, Magerle R, Krausch G: Self-assembly of functional nanostructures from ABC triblock copolymers. *Nat Mater* 2003, 2: 744–747.

61. Ren CL, Chen K, Ma YQ: Ordering mechanism of asymmetric diblock copolymers confined between polymer-grafted surfaces. *J Chem Phys* 2005, 122: 154904.

62. Ren CL, Ma YQ: Reentrant ordering transition of asymmetric copolymer solution film confined between polymer-grafted surfaces. *Phys Rev E* 2005, 72: 051804.

63. Ji S, Liu C-C, Son JG, Gotrik K, Craig GSW, Gopalan P, Himpsel FJ, Char K, Nealey PF: Generalization of the use of random copolymers to control the wetting behavior of block copolymer films. *Macromolecules* 2008, 41: 9098–9103.

64. Mansky P, Liu Y, Huang E, Russell TP, Hawker CJ: Controlling polymer-surface interactions with random copolymer brushes. *Science* 1997, 275: 1458–1460.

65. Drolet F, Fredrickson GH: Combinatorial screening of complex block copolymer assembly with self-consistent field theory. *Phys Rev Lett*1999, 83: 4317.

66. Drolet F, Fredrickson GH: Optimizing chain bridging in complex block copolymers. *Macromolecules* 2001, 34: 5317.

67. Kawakatsu T: *Statistical Physics of Polymers: an Introduction*. Berlin, Heidelberg: Springer; 2004.

68. Aubouy M, Fredrickson GH, Pincus P, Raphael E: End-tethered chains in polymeric matrices. *Macromolecules* 1995, 28: 2979–2981.

69. Jung YS, Jung W, Tuller HL, Ross CA: Nanowire conductive polymer Gas sensor patterned using self-assembled block copolymer lithography. *Nano Lett* 2008, 8: 3776–3780.

70. Guo ZJ, Zhang GJ, Qiu F, Zhang HD, Yang YL, Shi AC: Discovering ordered phases of block copolymers: new results from a generic fourier-space approach. *Phys Rev Lett* 2008, 101: 028301.

14

PHOTOTHERMAL EFFECTS AND APPLICATIONS OF POLYDIMETHYLSILOXANE MEMBRANES WITH CARBON NANOPARTICLES

Reinher Pimentel-Domínguez [1,†], Amado M. Velázquez-Benítez [1,†], J. Rodrigo Vélez-Cordero [2,3,†], Mathieu Hautefeuille[4,†], Francisco Sánchez-Arévalo [1,†] and Juan Hernández-Cordero [1,*,†]

[1] Instituto de Investigaciones en Materiales, Universidad Nacional Autónoma de México, A.P. 70-360, México City 04510, México

[2] Instituto de Física, Universidad Autónoma de San Luis Potosí, San Luis Potosí, S.L.P. 78290, México

[3] Cátedras CONACyT, Dirección Adjunta de Desarrollo Científico del CONACyT, México City 03940, México

[4] Facultad de Ciencias, Depto. de Física, UNAM, México City 04510, México

ABSTRACT

The advent of nanotechnology has triggered novel developments and applications for polymer-based membranes with embedded or coated nanoparticles. As an example, interaction of laser radiation with metallic and carbon nanoparticles has shown to provide optically triggered responses in otherwise transparent media. Incorporation of these materials inside polymers has led to generation of plasmonic and photothermal effects through the enhanced optical absorption of these polymer composites. In this work, we focus on the photothermal effects produced in polydimethylsiloxane (PDMS) membranes with embedded carbon nanoparticles via light absorption. Relevant physical parameters of these composites, such as nanoparticle concentration, density, geometry and dimensions, are used to analyze the photothermal features of the membranes. In particular, we analyze the heat generation and conduction in the membranes, showing that different effects can be achieved and controlled depending on the physical and thermal properties of the composite material. Several novel applications of these light responsive membranes are also demonstrated, including low-power laser-assisted micro-patterning and optomechanical deformation. Furthermore, we show that these polymer-nanoparticle composites can also be used as coatings in photonic and microfluidic applications, thereby offering an attractive platform for developing light-activated photonic and optofluidic devices.

KEYWORDS

PDMS-nanocarbon composites; photothermal conversion; photomechanical actuation

1. INTRODUCTION

A wide variety of synthetic membranes have been developed for applications involving separation processes, leading to useful solutions for wastewater treatments and contaminant removal [1]. Research in polymer membranes has been particularly extensive owing to their simple processing and higher flexibility for hosting other materials [2]. Recent efforts have been devoted to producing reconfigurable materials, in particular, polymers with the ability to adjust specific properties in response to external stimuli such as temperature and electromagnetic radiation [1,2,3]. As an example, changes in the geometry and dipole moment of photoresponsive molecules, such as azobenzenes and triphenylmethane, have proven useful for developing polymer membranes whose wettability and permeability features can be adjusted with light [4]. Other guests commonly used in host polymers include metal nanoparticles, which have shown to yield improved performance of membranes for wastewater treatment [1,2,3]. In general, the inclusion of nanoparticles in polymer membranes aims at improving features such as hydrophobicity and mechanical strength, although new functionalities may also be obtained upon properly combining a polymer host with adequate nanoparticles as guests. Nominally transparent polymers, for instance, may be used as hosts for optically absorbing nanoparticles, thereby providing a new platform for developing photoresponsive polymers.

An interesting photoresponsive material is obtained upon combining polydimethylsiloxane (PDMS) with carbon nanostructures. PDMS has been widely used due to its remarkable thermal, rheological, mechanical and biological properties [5,6,7]. The inclusion of carbon nanoparticles and nanotubes in PDMS membranes has led to obtaining photothermal effects, owing to the enhanced optical absorption observed in carbon-based nanostructures and the subsequent heat dissipation within the host polymer matrix [8,9,10,11]. When irradiated with low power laser light, either focused or guided by an optical fiber, these PDMS-carbon composites show highly localized thermal effects leading to applications such as laser-assisted microfabrication and optomechanical deformation [12,13,14]. This capability for generating highly localized thermal effects can be further exploited in photonics for fabrication, activation, and control of a wide variety of devices. Indeed, recent developments in coating deposition techniques for fiber optic devices have demonstrated that optically controlled tunable devices can be fabricated with PDMS-carbon composites, offering an attractive alternative to electrically activated materials [15]. These coatings can also be used on glass capillaries thereby providing a means to generate optically triggered thermocapillary flow [16], offering novel possibilities for developing optofluidic devices [17].

In this paper, we evaluate the performance of PDMS–carbon composites as photothermal materials for developing polymer membranes and coatings. We focus on their fabrication procedure, and, in particular, on the effects of nanoparticle distribution on their performance as photothermal materials. The composites are based on a PDMS matrix used as a host for carbon nanopowder; these two are mixed following simple procedures yielding membrane samples with different features. Their photothermal properties are evaluated by means of thermal and spectroscopic analyses, showing that the preparation procedure is relevant for obtaining an adequate temperature increase upon irradiating the samples with low power IR laser light. The temperature as a function of laser power is evaluated experimentally and then compared with a theoretical model incorporating the relevant thermal and optical parameters for the PDMS–carbon composites. Heat generation and transfer are evaluated for PDMS samples with embedded carbon nanopowder; furthermore, we also analyze PDMS membranes with a thin layer of carbon nanopowder evenly spread on its surface. While the former composites yield samples with photomechanical features, the irradiated surface of the samples with superficial carbon nanopowder shows a significant increase in temperature adequate for laser patterning. We provide evidence of both phenomena upon direct measurement of the stress induced on the membranes by laser irradiation, and through direct laser patterning on PDMS samples. Finally, we demonstrate a novel method using laser light for inducing two-phase flow in glass capillary tubes coated with the photoresponsive PDMS–carbon composite. We begin by analyzing the mechanisms for heat generation and conduction in polymer-carbon nanoparticle composites, considering relevant parameters such as absorption and thermal conductivity.

2. MATERIALS AND METHODS

2.1. Photothermal Effects in Polymer–Carbon Nanoparticle Composites

Composites such as PDMS with embedded carbon nanoparticles (PDMS/CNPs) fall into the category of what is known as photoresponsive or participating media [9,18,19]. The study of heat generation and transfer in such media constitutes a fascinating subject due in part to its implicit complexity. First of all, the dispersed phase in the composite behaves as scattering centers that deviate the incident radiation in different directions; hence, we are forced to solve not only the global energy equation but also the radiative heat equation, which has an integro-differential form [20,21]. Secondly, the heat transfer and generation parameters, such as the thermal diffusivity and absorption coefficient, change with temperature or are prone to the particular morphology of the dispersion [8]. This necessarily leads to situations where a single CNP concentration may yield distinct heat transfer characteristics depending on how the dispersion was made, or on the functionalization of the CNPs (if any). Despite these and other subtleties found in these kinds of composites, in this section, we intend to enlist some general guidelines governing the heat transfer properties in such systems. In fact, the optical properties of CNPs allow us to employ handy mathematical models for such a highly absorbing media. Indeed, for certain wavelengths and optical intensities, the absorption of PDMS/CNPs composites can be maximized with respect to the scattering effects, reducing the mathematical complexity by decoupling both governing equations leaving the heat transfer process described in terms of a conductive model.

A simple model to obtain the temperature profile across a membrane (thickness well above the mean free path of the energy carriers) is for the case when the composite is considered as a single continuous material, with an effective absorption coefficient β and thermal conductivity k [22]. We therefore assume that β and k include the dependence of the two-components properties of the material such as the CNP network morphology, dispersion size, CNP chemical modification and CNPs-PDMS heat interfacial resistance [8,23,24]. In addition, we also assume that the interrogating distances are greater than the mean inter-particle distance. A mathematical model with such assumptions can thus be set as follows [9]: consider a membrane with thickness L (see Figure 1) having effective β and k parameters and surrounded by dissipating media on both faces (heat transfer coefficients h_1 and h_2). In such a geometry (radial distances much higher than axial distances, *i.e.*, $0 \leq r \leq \infty$, $0 \leq z \leq L$), the increase in temperature ΔT can be computed for a time t using Green's functions in cylindrical coordinates:

$$\Delta T(r,z,t) = \frac{\alpha}{k}\int_{\tau=0}^{t}\int_{0}^{\infty}\int_{0}^{L} \mathcal{G}(r,z,t|r\prime,z\prime,\tau)\, q(r',z\prime)\, r'dz'dr'd\tau, \quad (1)$$

where

$$\mathcal{G}(r,z,t|r',z',\tau) = \frac{1}{\alpha(t-\tau)} exp\left[-\frac{r^2+r^{2'}}{4\alpha(t-\tau)}\right]\mathcal{I}_o\left[\frac{rr'}{2\alpha(t-\tau)}\right]\times$$
$$\sum_{n=1}^{\infty}\frac{e^{-\alpha\lambda_n^2(t-\tau)}}{N(\lambda_n)}\left[\lambda_n\cos(\lambda_n z)+H_1\sin(\lambda_n z)\right]\left[\lambda_n\cos(\lambda_n z')+H_1\sin(\lambda_n z')\right], \quad (2)$$

$$N(\lambda_n) = \left(\lambda_n^2+H_1^2\right)\left(L+\frac{H_2}{\lambda_n^2+H_2^2}\right)+H_1, \quad (3)$$

and the values of $\lambda n = \eta n/L$ are given by the transcendental function:

$$\tan(\eta_n) = \frac{\eta_n(Bi_1+Bi_2)}{\eta_n^2 - Bi_1 Bi_2} \text{ for n} = 1,2,3,\ldots \quad (4)$$

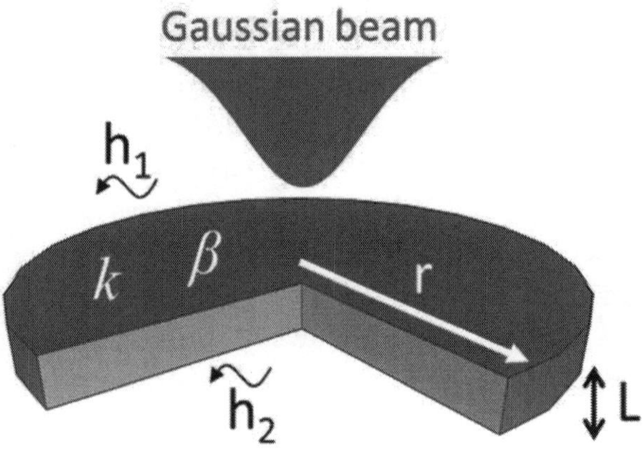

Figure 1. Schematic representation of a composite membrane showing the relevant parameters involved in the mathematical model describing light absorption and heat generation in PDMS/CNP composites (see text).

In the above equations, α is the effective thermal diffusivity, H1,2=h1,2/k is the ratio between the heat transfer coefficients and the thermal conductivity, and Bi1,2=H1,2 L is the Biot number. Subscripts 1, 2 correspond to the irradiated and rear surface, respectively. Let us consider two forms for the heat generation term q(r,z) depending on the procedure followed to incorporate the CNPs in the PDMS matrix. We consider as well that heat is generated by the photothermal response of the composite upon irradiation with a laser beam having Gaussian intensity profile (see Figure 1). For CNPs evenly dispersed within the polymeric matrix, the q(r,z) term can be formulated using Beer-Lambert's law [19,20]:

$$q\left(r,z\right) = \left(1-\mathcal{R}\right)\beta I_o \exp\left(-\beta z - r^2/A\right), \tag{5}$$

where A=w2/2 and *w* is the laser beam radius at which the optical intensity Io=P0/Aπ decays to e−2; P0 is the optical power and R accounts for the non-absorbed incident radiation (e.g., due to surface reflectivity). If the CNPs are deposited on the surface of one side of the membrane (as in the procedure where PDMS is cured over a bed of CNP, see below) the pertinent form of the generation term becomes:

$$q\left(r,z\right) = 2\left(1-\mathcal{R}\right) I_o \delta\left(z\right) \exp\left(-r^2/A\right), \tag{6}$$

This accounts for the fact that the radiation is absorbed on a very thin layer of material, *i.e.,*:

$$lim_{\beta\to\infty}\frac{\beta}{2}e^{-\beta z} = \delta\left(z\right), \tag{7}$$

where δ(z) is the delta function. Using this formulation and some integral formulas [9], we can obtain analytical solutions for ΔT at points of interest. For example, the transient ΔT at the point located at the rear surface exactly opposing the beam axis (*r* = 0) impinging on a membrane in which CNP are mixed with the PDMS has the form [9]:

$$\Delta T\left(L,t\right)_{axis}=\left(1-\mathcal{R}\right)\frac{A\beta I_o}{2k}\sum_{n=1}^{\infty}\frac{I_1\left(\lambda_n\right)}{N\left(\lambda_n\right)}\left[\lambda_n\cos\left(\lambda_n L\right)+H_1\sin\left(\lambda_n L\right)\right]e^{\frac{\lambda_n^2 A}{4}}$$
$$\times\left[Ei\left(-\alpha\lambda_n^2 t-\frac{A\lambda_n^2}{4}\right)-Ei\left(-\frac{A\lambda_n^2}{4}\right)\right]. \tag{8}$$

Here, Ei denotes the exponential integral and

$$I_1\left(\lambda_n\right)=\frac{1}{\beta^2+\lambda_n^2}\left[\lambda_n\left(\beta+H_1\right)+e^{-\beta L}\left\{\sin\left(\lambda_n L\right)\left(\lambda_n^2-H_1\beta\right)-\lambda_n\cos\left(\lambda_n L\right)\left(\beta+H_1\right)\right\}\right].$$

Similarly, the final expression for the case when CNP cover only the irradiated surface is:

$$\Delta T\left(L,t\right)_{axis}=\left(1-\mathcal{R}\right)\frac{AI_o}{k}\sum_{n=1}^{\infty}\frac{\lambda_n}{N\left(\lambda_n\right)}\left[\lambda_n\cos\left(\lambda_n L\right)+H_1\sin\left(\lambda_n L\right)\right]e^{\frac{\lambda_n^2 A}{4}}\times$$
$$\left[Ei\left(-\alpha\lambda_n^2 t-A\lambda_n^2/4\right)-Ei\left(-A\lambda_n^2/4\right)\right]. \tag{10}$$

Equations (8) and (10) show that ΔT maintain a linear relation with the optical power (Po~AIo). Hence, as far as the conduction regime is concerned, ΔT can be predicted for any value of Io once the factor ΔT/Io(k,β,h1,2,L,R,w) is known. Another interesting feature of the conduction regime is that ΔT is inversely proportional to the thermal conductivity *k*. Thus, membranes with good thermal conductivities will behave more like heat removers rather than heat generators (*i.e.*, they will oppose local increases of ΔT). Note also that ΔT increases with the absorption coefficient β, which is an expected behavior, although much less is known about how β depends on the CNP dispersion and/or clustering. Remarkably, the theoretical model predicts that the composites can be designed with different heat generation/dissipation properties due to the opposing dependencies of β and *k* on the dispersion morphology. The effects of CNP clustering on the thermal features of the composites will be explored in the following sections.

A final note should be given concerning the heat transfer coefficients h1,2 and their roles when comparing theoretical and experimental results. If the temperature is measured by non-contact devices (e.g., with an IR camera), h1,2 are easily known since they must correspond to the value of heat convection in air (~O(10) W/m2K); however, if ΔT is measured with contact devices (thermocouples, resistive temperature sensors), h2 will be larger, and usually unknown because direct contact with the membrane adds an additional dissipation and interfacial resistance [9].

2.2. PDMS/CNP Membranes: Materials and Preparation

Membranes of PDMS/CNP composites were fabricated using Dow Corning Sylgard 184 PDMS, with carbon nanopowder (particle size < 100 nm, Sigma Aldrich, St. Louis, MO, USA, CAS: 633100). The use of nanopowder instead of particles with defined geometries allows for minimizing any optical and/or thermal effect attributable to particle geometry. Furthermore, materials with high commercial availability were preferred in order to obtain low-cost materials for fabricating the membranes. Incorporation of the CNP is of crucial importance to generate a uniform distribution in order to yield homogeneous properties along the whole membrane. Two approaches were used when mixing the CNP with the PDMS aiming at obtaining different CNP cluster sizes. We first used a dispersing device (Ultra-Turrax T18, IKA Werke GmbH & Co., Staufen, Germany) for mixing both components of the composite membranes and these were labeled as PDMS+CNP. The second approach involved the use of chloroform during the mixing process, aiming only at reducing the viscosity of the silicone oil of the PDMS. Under this condition, the CNPs are easily incorporated into the polymer matrix by means of simple mixing using a magnetically driven stirrer, avoiding further modification of the materials [10,11,25]. The addition of $CHCl_3$yielded a homogeneous aqueous solution and the composites obtained with this procedure were labeled as PDMS+CNP+$CHCl_3$. The fabrication process is illustrated in Figure 2 and involves the following steps:

(a) The required base material of the elastomer (silicone oil) is weighted inside of a glass beaker.

(b) CNP is added in a concentration of 0.1% in weight compared to the total PDMS amount. For the PDMS+CNP mixture, the nanopowder is simply poured into the beaker; alternatively, chloroform is also added in a proportion of 0.75 mL per g of PDMS in order to obtain the mixture for the PDMS+CNP+$CHCl_3$ samples.

(c) The materials are mixed to evenly distribute the carbon nanopowder in the silicone oil. For the PDMS+CNP, mixing of the materials is achieved with the dispersing device operating for 5 min at a speed of 2800 rpm, and subsequently increasing the speed to 3600 rpm for 8 additional minutes. The second mixture, PDMS+CNP+$CHCl_3$, yields a liquid solution due to the addition of chloroform; this is mixed using a magnetic bar activated by a hot plate with magnetic stirrer while heated at 150 °C until full evaporation of the $CHCl_3$.

(d) The curing agent is added with a 1:10 ratio compared to the base material and both parts are mixed by hand during two minutes.

(e) Air captured during the mixing process is removed thru a degassing process.

(f) The PDMS/CNP mixture is poured into a glass mold; this was fabricated from glass substrates and height spacers of approximately 420 μm. The excess of material is then removed using the doctor blade technique.

(g) Solidification of the composites is achieved upon heating at 80 °C during 2 h.

(h) After cooling at room temperature, the resulting PDMS/CNP composite is finally removed from the mold.

Optical microscopy images of the resulting PDMS/CNP membranes are shown in Figure 3; for comparison, we also include an image of a PDMS membrane (*i.e.*, without CNP) cured under the same conditions as those followed for the composites. The PDMS+CNP samples show more uniformly distributed CNP within the membrane with only a few carbon clusters separated from each other (Figure 3b). In contrast, a larger amount of clusters in close proximity are apparent in the PDMS+CNP+$CHCl_3$ samples (see Figure 3c), owing to a less uniform distribution of CNP forming a percolating network morphology. The membranes obtained with the dispersing device (PDMS+CNP) thus show a more homogeneous distribution of CNP, and with much fewer clusters within the PDMS matrix. As demonstrated in the following sections, these features determine the photothermal performance of the composite membranes.

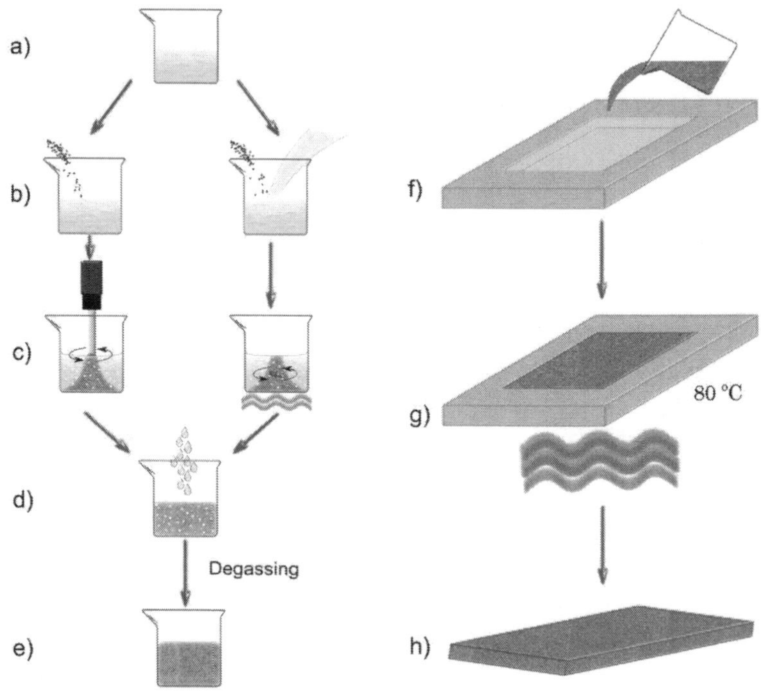

Figure 2. Preparation method for the PDMS/CNP membranes showing two mixing processes to obtain different CNP cluster sizes. (**a**) Weighting of the silicone oil inside of a glass beaker, (**b**) addition of the CNP to the PDMS, (**c**) mixing of the CNP with PDMS (evaporation by heat (orange lines) when $CHCl_3$ is used), (**d**) addition of curing reagent, (**e**) degassing of the blends, (**f**) PDMS/CNP mixture poured into a glass mold, (**g**) curing process by heat (orange lines) at 80 °C for solidification, and (**h**) unmolding PDMS/CNP samples.

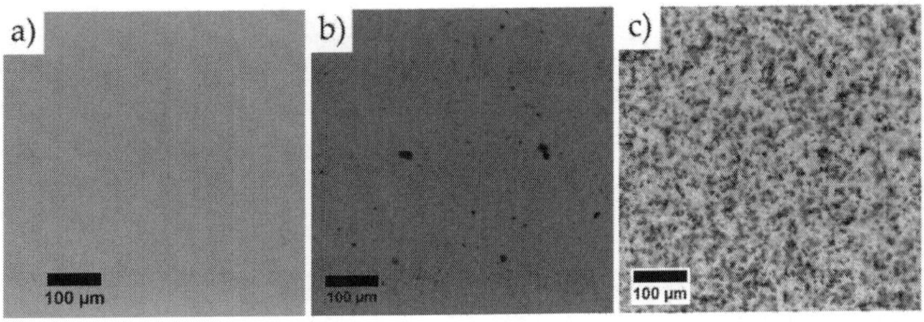

Figure 3. Optical microscopy images of the fabricated membranes: (**a**) pure PDMS; (**b**) PDMS+CNP; and (**c**) PDMS+CNP+$CHCl_3$. The magnification is the same for the three images (the scale bar is shown in (**a**)).

3. RESULTS AND DISCUSSION

Several tests were carried out to characterize the composite membranes. We focused on evaluating the effects of the fabrication process on the photothermal, structural and photomechanical features of the membranes.

3.1. Photothermal Effects: Heat Generation and Transfer in the Composites

We begin by analyzing the heat generation and transfer features as predicted by the mathematical model obtained inSection 2.1. Figure 4a shows the experimental and theoretical values of ΔT at the point ($r = 0$, $z = L$) for the PDMS+CNP and PDMS+CNP+$CHCl_3$ samples. Heat was generated in the membranes by direct irradiation with a 975 nm laser diode coupled to an optical fiber (Corning SMF-28e, Corning, NY, USA, numerical aperture NA = 0.14); the resulting output beam had a Gaussian profile and was located 5.82 mm away from the membranes. The experimental values of ΔT were measured using an IR camera (Fluke Ti300, Everett, WA, USA), while the theoretical values were obtained using Equation (8) and the parameters listed in Table 1.

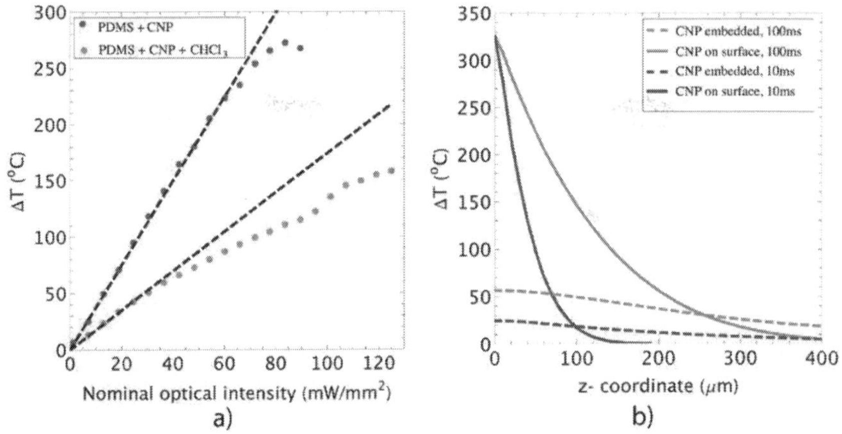

Figure 4. Heat generation and transfer in PDMS/CNP membranes: (**a**) comparison of the experimental ΔT (dots) and theoretical (dashed lines) values using Equation (8) and the parameters given in Table 1. The parameter n in Equation (8) was set to 48 and$t$ = 1 min, which was the experimental time to attain a steady value. The nominal optical intensities were calculated as IN=Po/πw2o , where w_o is the spot radius estimated by using the NA of the optical fiber and fiber tip-membrane separation (5.82 mm); (**b**) comparison of the theoretical ΔT profiles along the membrane thickness using Equation (8) and (10) for the two CNP dispersion types: one where the CNP are dispersed in all the polymer matrix, and another where the CNP are deposited just on the irradiated surface.

The close agreement between the experimental measurements and the theoretical values at low optical powers confirm that ΔT is linearly proportional to the optical intensity, thus validating the conduction regime assumption used in the model. Notably, Figure 4a suggests that deviation from the linear response most likely arises from changes in the optical properties of the material (e.g., reflectivity or emissivity at the surfaces) rather than on parameters subject to temperature changes. Another interesting observation is that the linear-conduction regime holds even though optical limiting behaviors (scattering effects) are already present in the membranes. It has been shown that optical limiting effects appear at values generally above 1 J/cm² for similar wavelengths as those employed in this work [21,26]; this value is easily overpassed in our experiments considering that the laser beam is maintained for periods of about one minute. A complete interpretation of such results will demand computation of the radiative heat flux inside the membrane. However, a simple phenomenological description can be given in terms of the absorption/scattering compensation: while scattering increases with the optical intensity IN along with the propagation of light inside the material, the ΔT generated by the newly illuminated volume is compensated by a reduction in the absorption coefficient [21] (recall that the extinction coefficient is the sum of the absorption and scattering terms).

Another important observation can be made regarding the values of β and k. The experimental measurements shown inTable 1 indicate that β increases for an improved dispersion of the CNP fillers: while $\beta = 4.4$ mm^{-1} for the membranes fabricated with the dispersing device, $\beta = 2.8$ mm^{-1} for those using chloroform. Notice, however, that fitting Equation (8) with the experimental data yields a decrease in k with the dispersion of CNP (see Table 1), confirming that β and k maintain opposite dependences on the composite morphology. As seen in Figure 3, the CNPs are more homogeneously distributed in the PDMS+CNP membranes. In contrast, the PDMS+CNP+$CHCl_3$ samples show a large number of clusters of CNP distributed within the sample. While an improved distribution of CNP leads to a larger attenuation coefficient (PDMS+CNP), clusters within the membranes lead to improved thermal conductivity, associated with a percolating network morphology (PDMS+CNP+$CHCl_3$). These results can thus be interpreted in terms of the CNP interconnectivity. The PDMS+CNP samples include only a few isolated clusters with low connectivity, yielding a membrane with good thermal generation capabilities but with poor thermal conductivity (good thermal accumulation, low thermal dissipation). On the other hand, the PDMS+CNP+$CHCl_3$ samples contain a larger amount of clusters with high interconnectivity, yielding a membrane with poor thermal accumulation but good dispersion qualities. As shown in the following sections, the CNP also increases the crystallinity compared to pristine PDMS. This feature, together with the improved CNP clusters interconnectivity obtained when using chloroform as solvent, provides a plausible explanation of the increase in thermal conductivity obtained in our experiments.

Table 1. Parameters used in Equation (8) to fit the experimental data shown in Figure 4.

Parameter	PDMS+CNP	PDMS+CNP+$CHCl_3$	Notes
L (μm)	480	480	As determined from direct measurements.
A (mm^2)	0.1465	0.1465	Estimated using the fiber tip-membrane separation distance.
β (mm^{-1})	4.4	2.8	Determined from experiments (see below).
$h_{1,2}$ ($W/(m^2 \cdot K)$)	28	28	Typical value of heat convection in air.
k ($W/(m \cdot K)$)	0.26	0.6	Typical value for pristine PDMS is 0.1678 at 14.7 °C and 12,500 cst [24].
α (mm^2/s)	0.18	0.42	Evaluated as $\alpha = k/\rho\, Cp$, where Cp is the specific heat capacity and ρ is the materials density.
$\mathcal{R}$	0.04	0.04	Estimated considering a refractive index of 1.5

We finish this section comparing the theoretical ΔT profiles inside the membranes made by mixing the CNP with the PDMS and the others having CNP spread at the irradiated surface (see Equation (10)). Figure 4b shows the resulting ΔTprofiles obtained along the axial direction using Equations (8) and (10). The parameters used in both equations were again taken from Table 1 from the PDMS+CNP membrane. Two situations have been included in the comparative plot; in one case, both membrane conformations are irradiated with a Gaussian beam (IN=73.5 mW/mm2) during 100 ms (*i.e.*, 0.73 J/cm^2); in the other case, the membranes are irradiated with IN=209.8 mW/mm2 during 10 ms (*i.e.*, 0.21 J/cm^2). Clearly, while the membranes with embedded CNP generate, at the most, 20 and 50 °C during these short time periods, the membranes having CNPs spread at the irradiated surface attain a maximum $\Delta T \sim 300$ °C at $z = 0$, while, on the rear surface ($z = L$), the temperature is barely increased. Such an increase in temperature corresponds to the flash point of PDMS [27] and allows for laser ablation of the composite. As discussed in the final sections, this feature allows for laser patterning of the composite membranes.

3.2. Structural and Mechanical Characterization of the Composites

Thermogravimetric analysis (TGA) provides information about the thermal decomposition of the samples as a function of temperature. The TGA results for samples of CNP, pure PDMS and the PDMS/CNP composites are shown in Figure 5a. Clearly, the three polymer-based samples are completely stable from room temperature (25 °C) and up to temperatures just below 200 °C; beyond this temperature, the thermal degradation of pure PDMS occurred at 350 °C. Notice, however, that the CNP start degrading at even lower temperatures. Upon analyzing the first derivatives of the TGA curves, two stages can be identified at which significant weight loss occurs (Figure 5b). The first one is located between 300 and 400 °C and corresponds to PDMS degradation into cyclic oligomers due to Si–O bond scission [28]. The second stage is observed when the PDMS is exposed to higher temperatures (480 and 626 °C) at a slow heating rate; hence, thermal degradation of PDMS starts via depolymerization of the polysiloxane backbone. This leads to the formation of cyclosiloxanes [29], and it has been shown that a radical mechanism occurs throughout the Si–CH_3 bond scission yielding methane through hydrogen abstraction [30]. As a consequence, the flexibility of PDMS decreases due to the crosslinking effect of macro-radicals; this, in turn, increases the thermal stability of the sample, and bond reorganization occurs leading to silicon–oxycarbide residues. The thermal degradation of PDMS can be affected by factors such as end-group functionality, impurities or solvent and oxidation. In our case, the addition of a small percentage of CNP (0.1%) slightly reduced the thermal degradation of the composite membranes compared to pristine PDMS samples (around 2%). Notice also that no significant differences are apparent for the thermal degradation of the two composite membranes, occurring between 250 and 300 °C.

X-ray diffraction analysis did not show significant carbon nanoparticle features, owing to the low concentration (0.1%*w/w*) used to fabricate the membranes (see Figure 5c). Nonetheless, the characteristic peaks of PDMS were easily detected, with the most intense peak located at 11.5° corresponding to its tetragonal phase. Only a small change in the diffraction pattern was observed for the composites, occurring at 26.5°, hardly perceptible, and possibly attributable to the inclusion of the CNP. The percentage of crystallinity for the samples was found to be 24% for PDMS, 38.3% for PDMS+CNP, and 38.4% for PDMS+CNP+$CHCl_3$. This improvement in crystallinity observed in both samples can be attributed to an arrangement of polymeric chains favored by the CNP acting as nucleation centers yielding crystalline zones. In principle, an increase in crystallinity should improve the thermal stability; however, in our experiments, the composite membranes show thermal degradation at temperatures close to those observed for pristine PDMS. Notice, however, that the CNPs incorporated in the membranes start degrading even at lower temperatures, thus contributing to the registered weight loss during the TGA test. Further weight loss is accounted for by PDMS degradation occurring within the same temperature range.

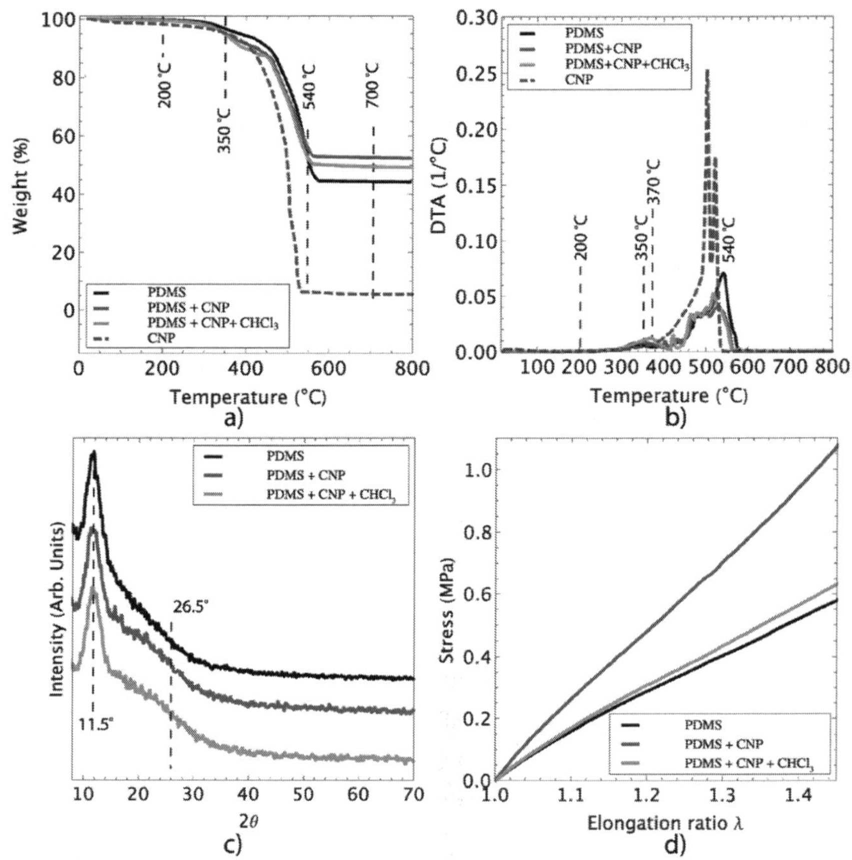

Figure 5. Thermal and mechanical characterization of the composite membranes: (**a**) thermogravimetric analysis of PDMS and its composites; (**b**) derivatives of thermogravimetric analysis evidencing the critical temperatures of degradation for PDMS and its composites; (**c**) X-ray diffraction spectra showing amorphous and crystalline features of PDMS and composites; (**d**) mechanical response of PDMS and composites under uniaxial tensile load.

To evaluate the mechanical behavior of the composites, dog-bone specimens were prepared following the ASTM D1708 standard. These were then evaluated by means of uniaxial tensile experiments, carried out with a custom-designed device [31]. As shown in Figure 5d, the registered stress for PDMS and its composites varies in a nonlinear fashion as a function of the elongation ratio (λ). This is consistent with previous reports on elastomeric composites with carbon-based nanoparticles; from the resulting curves of the uniaxial tests, the shear modulus and Young's modulus of the membranes can be calculated as reported elsewhere [14]. For comparison purposes, the stress curves shown in Figure 5d are plotted only up to $\lambda = 1.4$; notice that the inclusion of CNP clearly modifies the mechanical behavior of the samples. In particular, when compared with pure PDMS, the composite membranes show an increase in the shear module, as observed by the higher stresses registered for similar elongation ratios. Furthermore, the curves provide evidence of the influence of the fabrication process on the mechanical behavior of the membranes. The PDMS+CNP showed higher stresses owing to a more homogeneous distribution of CNP clusters within the PDMS matrix (see Figure 3). In contrast, the PDMS+CNP+$CHCl_3$ membranes showed only a slight improvement when compared to pure PDMS. Although the crystallinity of PDMS+CNP and PDMS+CNP+$CHCl_3$ were almost the same, the mechanical behavior of the composites was different. In particular, the shear modulus of PDMS+CNP was estimated to be 0.87 MPa, larger than that of the PDMS+CNP+$CHCl_3$ membrane estimated as 0.58 MPa. Meanwhile, the pristine PDMS membrane resulted in a 0.55 MPa shear module. These results suggest that the composite prepared by mechanical stirring at higher RPMs (Ultra-Turrax T18 mixing device, IKA Werke GmbH & Co.) effectively reduces the size of the carbon clusters (see Figure 3b). This in turn favors the incorporation of the carbon particles within the polymer network, thus reinforcing the polymer matrix. In contrast, the large number of CNP clusters observed in the PDMS+CNP+$CHCl_3$ membranes seems to be acting as inclusions favoring stress concentration without increasing the strength significantly.

3.3. Photomechanical Effects: Light Induced Stress

To evaluate the photomechanical response of the membranes, the samples were subjected to a stress relaxation test and irradiated with laser light. The test was performed for elongation ratios of $1.4 < \lambda < 1.7$, and laser irradiation was delivered via the infrared (IR) laser diode. Figure 6a shows the stress relaxation curves registered for the samples during the test; noticeably, laser irradiation induces an increase in stress in the samples. Further evidence of this effect is provided in Figure 6b, showing the stress registered in each sample for five pulses of the IR laser (optical power of 136 mW). This effect induced by laser light is due to optical absorption of the CNP and subsequent heat release within the polymer matrix [11,12,13,14]. The inclusion of CNP in PDMS thus provides a means to obtain photothermal effects in an otherwise transparent material. The increase in stress in the samples is explained upon considering the PDMS and its composites as systems whose entropy is reduced when they are pre-stretched in the relaxation test. Pre-stretching causes a reduction in entropy because the polymeric chains tend to align along the direction of the applied force. Hence, when the IR laser is turned on, and after some characteristic time (20 s for the tested samples), the photothermal effect generates a highly localized increase in temperature yielding heat percolation within the polymeric matrix. This leads to an increase in entropy and the polymeric chains that were previously aligned during the pre-stretching recover their initial state. Globally, this set of events produce mechanical contraction of the sample, and this is registered by the load cell and graphically represented by a peak in the stress relaxation curve (see Figure 6b). Similar behavior has been observed in graphene/elastomer composites [32,33] and PDMS doped with carbon soot particles [14].

Measurements of the extinction coefficient (Figure 6c) clearly show that pristine PDMS is optically transparent to IR irradiation, while PDMS+CNP and PDMS+CNP+$CHCl_3$ indeed show extinction coefficients arising from absorption and scattering due to the inclusion of CNP. Notice that the membranes prepared with $CHCl_3$ had a lower extinction coefficient compared to that obtained for the membranes prepared with the dispersing device. In contrast, as shown in Table 1, the use of $CHCl_3$ during membrane fabrication yields larger heat conduction coefficients and a smaller temperature increase compared with those attained with the other membranes. As shown in Figure 6d, the fabrication method also affects the photomechanical response of the membranes, *i.e.*, the stress induced by laser irradiation. While both samples with CNP showed a five-fold increase in this response compared to pristine PDMS, the membranes with $CHCl_3$ yielded the largest stress under laser irradiation perhaps due to the presence of larger carbon clusters. Therefore, albeit achieving a lower increase in temperature, the improved heat conduction in the $CHCl_3$ membranes seems to favor the photomechanical response. Clearly, different effects related to the photothermal phenomenon can be achieved depending on the fabrication procedure. In what follows, we demonstrate a few examples showing potential applications for these laser driven membranes.

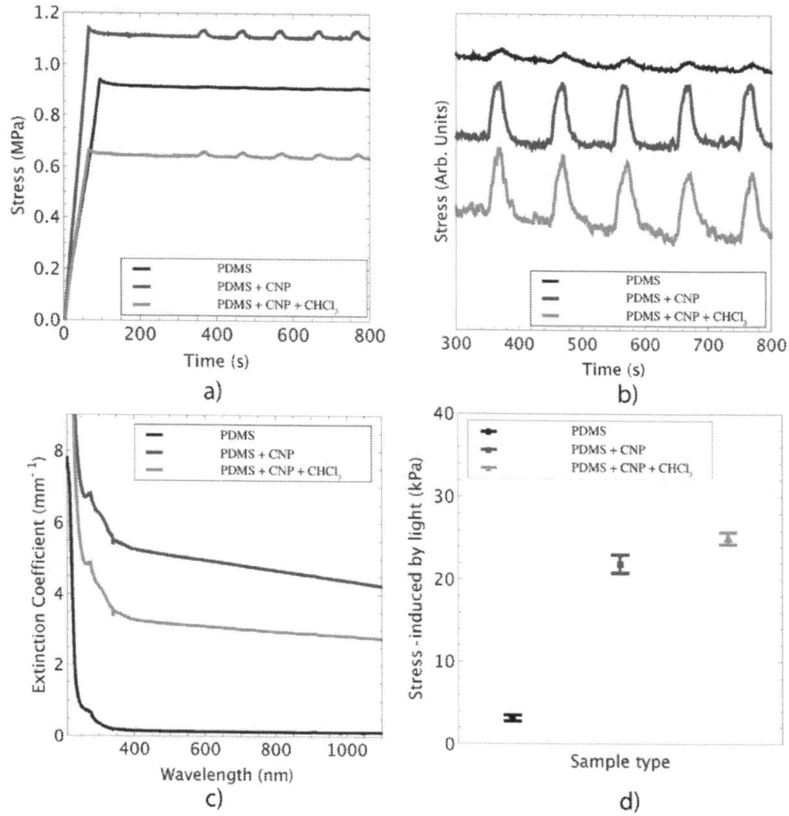

Figure 6. Laser induced mechanical stress in PDMS composites. (**a**) experimental curves showing the stress relaxation as function of time for PDMS and its composites; (**b**) peaks evidencing the photo-actuation capabilities of these composites; (**c**) extinction coefficient as a function of wavelength for PDMS and composites; and (**d**) photomechanical actuation of PDMS and composites.

3.4. Applications

The PDMS/CNP blend can be applied and cured in different manners such as thin films, coatings, or even using mold replica techniques to obtain diverse geometries. In the following subsections, we will demonstrate a few devices created with this composite, showing the versatility of this photothermal material and its usefulness to provide light activated functionalities and novel alternatives for fabrication of optically reconfigurable devices.

3.4.1. Coatings for Capillaries

Since the advent of soft lithography techniques, it has long been demonstrated that PDMS, together with other kinds of polymers, constitute hallmark materials for microfluidic applications [34,35]. One of the main reasons for this is because polymers can be easily molded and incorporated in a wide variety of microfluidic designs. Furthermore, these polymers can be used as matrices to host other types of materials such as CNP. This combination adds functionalities to these kinds of composites, allowing for the development of so-called smart micro-devices. PDMS/CNP composites can also be deposited along glass capillaries using fiber coating techniques, allowing for fabrication of photo-thermo-responsive devices capable of controlling thermocapillary flow [16]. This capability can lead to the realization of opto-pneumatic systems functioning as light-driven pistons for pumping fluids into microfluidic devices [17]. In this section, we demonstrate a new and straightforward application of PDMS/CNP coatings, used to generate two-phase boiling flows in micro-channels.

Boiling flows in micro-channels has been an interesting subject because they have helped in understanding prototypes of cooling systems for micro-electronic circuits [36]. The basic operational idea of such systems can be stated as follows: consider a micro-channel with a flowing liquid intended to withdraw heat from a high temperature source. For a given configuration, an optimum liquid flow rate has to be obtained capable of providing maximum heat dissipation while simultaneously keeping optimal flow conditions in the micro-channel.

At intermediate flow rates, heat is dissipated by the liquid but also by phase transitions, since part of the liquid evaporates in the form of micro-bubbles (heat removal is increased thanks to the latent heat of vaporization). Experimentally, we produced two-phase flows in a micro-channel by evaporating water inside a glass capillary (O.D. 330 μm, I.D. 200 μm) using a PDMS+CNP coating irradiated with laser light for heating (see Figure 7a). Figure 7b shows that different flow regimes can be produced, including the optimal bubbly flow regime. These were achieved upon adjusting the flow rate inside the capillary (up to 3.5 mL/h, or 30 mm/s) and the optical intensity at the output of the optical fiber (8 to 16 W/mm²). We are currently rationalizing these results in terms of two-phase boiling models and put this work in perspective with others that have employed different experimental setups.

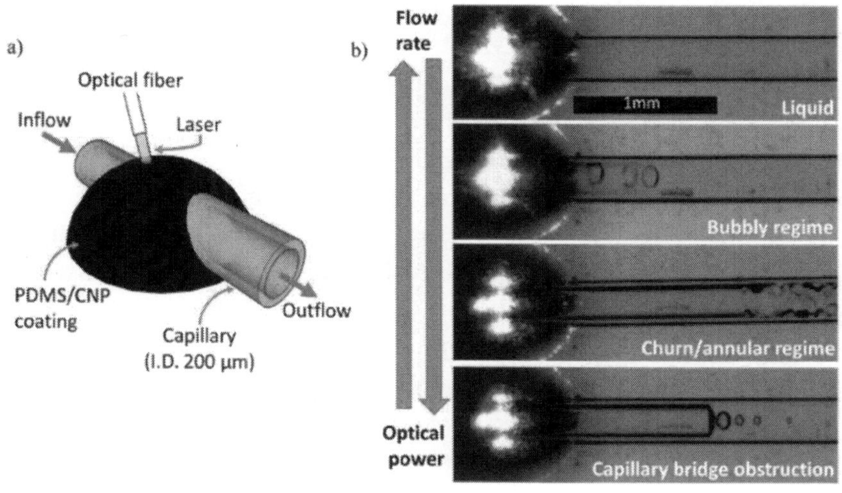

Figure 7. CNP/PDMS coatings for boiling flows: (**a**) schematic of the basic components to produce a boiling flow using PDMS+CNP coatings; (**b**) images showing the different flow regimes that can be achieved inside the capillary; the gas-liquid mixture can be varied by adjustments on the flow rate or the optical intensity of the laser diode. The bubbly flow regime constitutes the optimal flow condition since it allows good heat removal and optimal flow conditions. High optical powers or very low flow rates can create gas chambers or capillary bridges than can obstruct or introduce variations in the main flow.

3.4.2. Coating for Fiber Optic Devices

Photothermal effects at the micrometric scale provide a useful tool for developing light-activated devices based on optical fibers. Coating of optical fibers has been shown to be a viable approach to generate light interaction with thin films of different materials. In particular, thin film deposition on optical fiber tips has been used to generate ultrasound probes [37,38]. Coatings of composite materials may also be deposited on intermediate sections of the fibers yielding light driven tunable devices [39]. Heat generation and detection in specific sections of an optical fiber is also of interest for optical sensors and their incorporation in multifunctional systems.

To demonstrate the potential of PDMS/CNP composites for realizing optically tunable fiber devices, a thin film coating (approximately 1.7 μm thick) of the PDMS+CNP+$CHCl_3$ composite was applied on a fiber Bragg Grating (FBG) using the system described in [15]. After curing at 80 °C for 2 h, the coated FBG was irradiated with the 975 nm laser diode focused with a cylindrical lens to obtain uniform illumination over a length of 3 mm (see Figure 8a). The reflection peak of the FBG spectrum was initially centered at a wavelength of 1535.54 nm, and this was monitored with an optical spectrum analyzer as the output power of the laser diode was increased. As shown in Figure 8b, a maximum wavelength shift of approximately 170 pm was registered for the range of optical intensities used in this experiment. Using the typical value for the temperature sensitivity of FBGs (~10 nm/°C [40]), the resulting shift in peak wavelength is equivalent to a local increase in temperature of 16.8 °C. Heating of the FBG when irradiated with the laser was further confirmed with the thermographic camera (see inset in Figure 8b), showing that the temperature is effectively increased only within the illuminated region. Hence, laser tunable FBGs can be attained with the proposed approach offering a novel means for realizing tunable fiber optic filters as an alternative to devices activated by electro-optic effects.

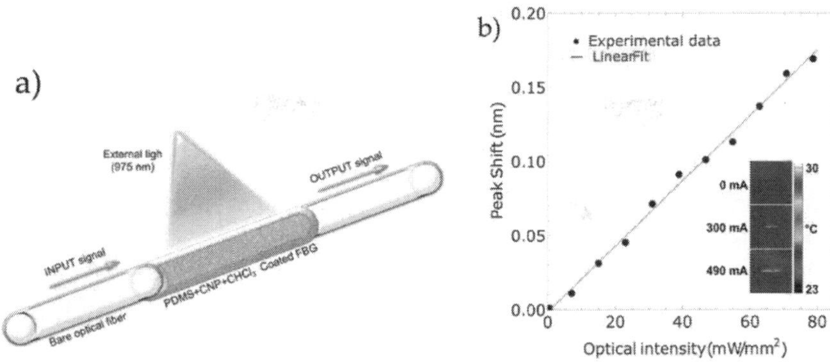

Figure 8. (**a**) schematic of FBG coated with a thin layer of the PDMS+CNP+$CHCl_3$composite irradiated with a focused IR laser beam; (**b**) FBG peak shift generated by light irradiation of the composite. The inset shows images obtained with the thermographic camera demonstrating localized heating on the irradiated section.

3.4.3. Laser Micropatterning

As shown elsewhere, CNP-coated PDMS membranes may be etched using a low power, near infrared laser from a CD pickup head [12]. A motorized computer numerical control setup allows for tightly focusing the laser beam at the surface of the modified PDMS sample to ablate the desired micropatterns inside the material. This is achieved in a local fashion using laser-induced incandescence. However, as observed experimentally, this only happens when the CNPs coat the PDMS layer with a certain adhesion to the surface; otherwise, particles are immediately propelled away from the laser focal spot when light is shone either in CW or pulsed mode [12]. Features such as the dimensions of the obtained patterns depend strongly on the laser parameters. However, thickness, homogeneity and roughness of the CNP layer also play an important role.

In general, PDMS membranes with embedded CNP are desirable because flat surfaces yield improved resolution for the patterning process. However, for these membranes, our direct laser etching technique has proven to be inefficient, as no laser-induced incandescence, originating the ablation, had been obtained before. This is consistent with our findings since significant differences in heating and temperature increase are expected if the carbon nanoparticles are embedded inside the polymer or only coating the surface of the PDMS (see Section 3.1). When the CNP are embedded in PDMS, the generated heat is driven away from the focal spot and hence neither incandescence nor combustion occur.

Upon combining embedment and coating of CNP in PDMS membranes, we have finally been able to confirm that a thin, superficial but embedded layer would allow reaching high temperatures above the combustion threshold. Using a newly adapted simple method described in Figure 9, thin layers of carbon nanoparticles were embedded immediately underneath the surface of interest. A doctor blading technique was used to coat a flat surface of a clean, hard substrate (typically low roughness acrylic sheets or glass slide) with a thin layer of CNP in alcohol. Then, a small volume of PDMS was drop cast on this layer and cured immediately at 80 °C for 2 h. Spin coating may be used for thinner layers with similar results. As shown in Figure 9, after detaching the cured membrane, the surface in contact with the CNP layer replicated the flatness of the hard substrate and no additional roughness due to the CNP was observed on our new membranes.

Unlike our previous PDMS/CNP samples, the process used to fabricate our new membranes allowed for etching the surface of the PDMS layer with improved resolution when compared with our previous results (see Figure 10). An optimal 2 μm wide spot was obtained in the best case, which is better than what was previously obtained with a CNP coating layer on PDMS [12]. This may be due to the quality of the PDMS surface, which leads to reducing scattering effects; furthermore, the absence of CNP roughness and reduction in homogeneity allow for concentrating more light, thereby obtaining highly localized thermal effects. Indeed, a much greater CNP homogeneity has been observed in these samples when compared to what is typically obtained with direct PDMS coating [12], thus leading to a greater in-plane resolution. Moreover, the laser ablation process is similar to that previously reported, offering the same control over all for the important laser parameters. It is interesting to remark that the depth of the ablated patterns is typically lower than what is obtained with direct PDMS/CNP coatings, reaching a maximum of 10 μm in the best case, as observed with a KLA-Tencor D600 profilometer. In addition, the optically absorbing layer is fully removed after etching, offering a great optical contrast between transparent and opaque areas for the development of optical masks, improving our previously reported results [41]. However, one disadvantage of this technique compared to that presented in [12] is that it is impossible to wash the residual CNP after the etching process; this limits its use for samples that do not require optical transparency around laser-etched micropatterns.

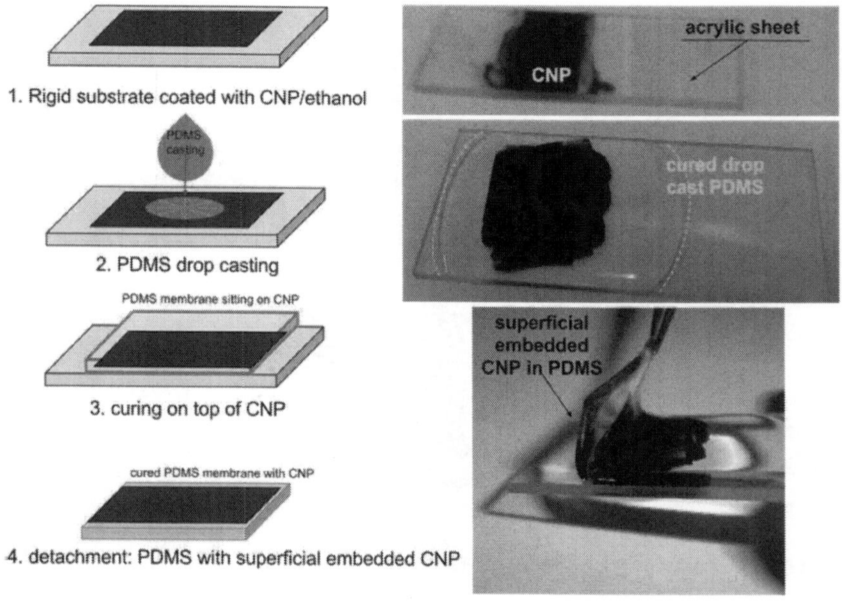

Figure 9. Adapted preparation method for superficially embedded CNPs in PDMS membranes.

4. CONCLUSIONS

We have analyzed the photothermal effects produced in PDMS membranes with embedded carbon nanoparticles via light absorption. A simple heat conduction model along with experimental analysis was used to analyze the photothermal features of the membranes. As confirmed experimentally, heat generation and conduction in the membranes allow for obtaining different effects, such as photomechanical actuation and incandescence, that can be controlled depending on the physical and thermal properties of the composite material. The PDMS/CNP composites offer attractive features, such as ease of fabrication and processing, both of which are attractive for applications involving photothermal materials. In particular, we demonstrated that the composites may be used as membranes for laser micropatterning, or may be incorporated as coatings for microfluidic and fiber optic devices as new and simple alternatives to already existing technologies and fabrication methods. Variations when preparing the PDMS/CNP composites allow for achieving different photothermal features, which are closely related to carbon cluster distribution. This versatility makes these materials an attractive platform for developing light activated photonic and optofluidic devices.

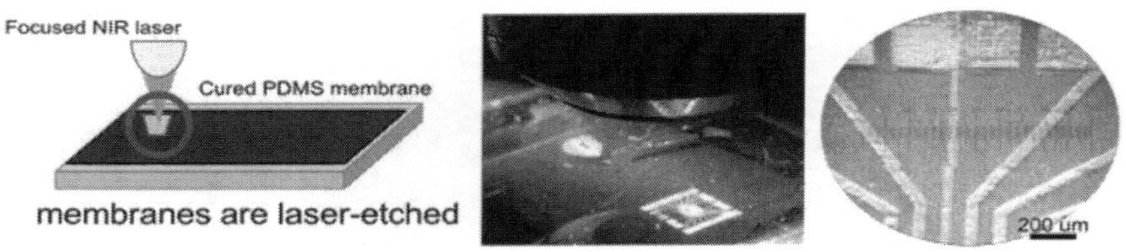

Figure 10. Example of a micropatterned design transferred to PDMS.

ACKNOWLEDGMENTS

This work was developed with financial support from PAPIIT, DGAPA-UNAM, through grants IT101215, IT100215 and TA100315. Support from Laboratorio Nacional de Soluciones Biomiméticas para Diagnóstico y Terapia LaNSBIODyT and CONACyT, grant 246988 is also greatly appreciated. The authors are grateful to Aarón Cruz Ramírez for his support in the laser etching process, Adriana Tejeda (Instituto de

Investigaciones en Materiales, Universidad Nacional Autónoma de México) for the X-ray measurements, and Damaris Cabrero (Instituto de Investigaciones en Materiales, Universidad Nacional Autónoma de México) for the TGA and DSC measurements. J. Rodrigo Vélez-Cordero acknowledges financial support from Cátedras Conacyt.

AUTHOR CONTRIBUTIONS

Reinher Pimentel-Domínguez and Amado M. Velázquez-Benítez manufactured the PDMS/CNP membranes and performed the coating and characterization of fiber optic devices. J. Rodrigo Vélez-Cordero carried out the theoretical characterization of heat generation and dissipation in the membranes, as well as two-phase flow generation experiments. Mathieu Hautefeuille performed the laser micropatterning experiments. Francisco Sánchez-Arévalo carried out the thermal, structural and mechanical characterization, as well as the evaluation of the photomechanical effects. Juan Hernández-Cordero did data analysis, experimental conception and design. All the authors contributed equally to writing the paper and with reagents, materials and analysis tools.

CONFLICTS OF INTEREST

The authors declare no conflict of interest.

REFERENCES

1. Kim, J.; van der Bruggen, B. The use of nanoparticles in polymeric and ceramic membrane structures: Review of manufacturing procedures and performance improvement for water treatment. *Environ. Pollut.* **2010**, *158*, 2335–2349.

2. Ng, L.Y.; Mohammad, A.W.; Leo, C.P.; Hilal, N. Polymeric membranes incorporated with metal/metal oxide particles: A comprehensive review. *Desalination* **2013**, *308*, 15–33.

3. Hu, L.; Gao, S.; Ding, X.; Wang, D.; Jiang, J.; Jin, J.; Jiang, L. Photothermo-responsive single-walled carbon nanotube-based ultrathin membranes for on/off switchable separation of oil-in-water nanoemulsions. *ACS Nano* **2015**, *9*, 4835–4842.

4. Nicoletta, F.P.; Cupelli, D.; Formoso, P.; De Filpo, G.; Colella, V.; Gugliuzza, A. Light responsive polymer membranes: A review. *Membranes* **2012**, *2*, 134–197.

5. Mata, A.; Fleischman, A.; Roy, S. Characterization of polydimethylsiloxane (PDMS) properties for biomedical micro/nanosystems. *Biomed. Microdevices* **2005**, *7*, 281–293.

6. Alrifail, A.; Lindahl, O.A.; Ramser, K. Polymer-based microfluidic devices for pharmacy, biology and tissue engineering. *Polymers* **2012**, *4*, 1349–1398.

7. Seethapathy, S.; Górecki, T. Applications of polydimethylsiloxane in analytical chemistry: A review. *Anal. Chim. Acta***2012**, *750*, 48–62.

8. Han, Z.; Fina, A. Thermal conductivity of carbon nanotubes and their polymer nanocomposites: A review. *Prog. Polym. Sci.* **2011**, *36*, 914–944.

9. Vélez-Cordero, J.R.; Hernández-Cordero, J. Heat generation and conduction in PDMS-carbon nanoparticle membranes irradiated with optical fibers. *Int. J. Therm. Sci.* **2015**, *96*, 12–22.

10. Miyako, E.; Nagata, H.; Hirano, K.; Hirotsu, T. Carbon nanotube–polymer composite for light-driven microthermal control. *Angew. Chem. Int. Ed.* **2008**, *47*, 3610–3613.

11. Miyako, E.; Nagata, H.; Hirano, K.; Hirotsu, T. Laser-triggered carbon nanotube microdevice for remote control of biocatalytic reactions. *Lab Chip* **2009**, *9*, 788–794.

12. Hautefeuille, M.; Cabriales, L.; Pimentel-Domínguez, R.; Velázquez, V.; Hernández-Cordero, J.; Oropeza-Ramos, L.; Rivera, M.; Carreón-Castro, M.P.; Grether, M.; López-Moreno, E. New perspectives for direct PDMS microfabrication using a CD-DVD laser. *Lab Chip* **2013**, *13*, 4848–4854.

13. Ahir, S.V.; Terentjev, E.M. Photomechanical actuation in polymer-nanotube composites. *Nat. Mater.* **2005**, *4*, 491–495.

14. Sánchez-Arévalo, F.M.; Garnica-Palafox, I.M.; Jagdale, P.; Hernández-Cordero, J.; Rodil, S.; Okonkwo, A.; Robles Hernandez, F.; Tagliaferro, A. Photomechanical response of composites based on PDMS and carbon soot nanoparticles under IR laser irradiation. *Opt. Mater. Express* **2015**, *5*, 1792–1805.

15. Velázquez-Benítez, A.M.; Reyes-Medrano, M.; Vélez-Cordero, J.R.; Hernández-Cordero, J. Controlled deposition of polymer coatings on cylindrical photonic devices. *J. Lightwave Technol.* **2015**, *33*, 176–182.

16. Vélez-Cordero, J.R.; Velázquez-Benítez, A.M.; Hernández-Cordero, J. Thermocapillary flow in glass tubes coated with photoresponsive layers. *Langmuir* **2014**, *30*, 5326–5336.

17. Vélez-Cordero, J.R.; Pérez Zúñiga, M.G.; Hernández-Cordero, J. An optopneumatic piston for microfluidics. *Lab Chip***2015**, *15*, 1335–1342.

18. Miyako, E.; Nagata, H.; Hirano, K.; Hirotsu, T. Micropatterned carbon nanotube-gel composite as photothermal material. *Adv. Mater.* **2009**, *21*, 2819–2823.

19. Tjahjono, I.K.; Bayazitoglu, Y. Near-infrared light heating of a slab by embedded nanoparticles. *Int. J. Heat Mass Transfer* **2008**, *51*, 1505–1515.

20. Modest, M.F. *Radiative Heat Transfer*, 3rd ed.; Academic Press: Amsterdam, The Netherlands, 2013.

21. Wang, J.; Chen, Y.; Li, R.; Dong, H.; Zhang, L.; Lotya, M.; Coleman, J.N.; Blau, W.J. Nonlinear optical properties of graphene and carbon nanotube composites. In *Carbon Nanotubes—Synthesis, Characterization, Applications*; Yellampalli, S., Ed.; InTech: Rijeka, Croatia, 2011.

22. Lax, M. Temperature rise induced by a laser beam. *J. Appl. Phys.* **1977**, *48*, 3919–3924.

23. Han, C.-W.; Birringer, R.; Clarke, D.R.; Gleiter, H. Effective thermal conductivity of particulate composites with interfacial thermal resistance. *J. Appl. Phys.* **1997**, *81*, 6692–6699.

24. Hong, J.; Lee, J.; Hong, C.-K.; Shim, S.-E. Effect of dispersion state of carbon nanotube on the thermal conductivity of poly(dimethyl siloxane) composites. *Curr. Appl. Phys.* **2010**, *10*, 359–363.

25. Majeed, S.; Filiz, V.; Shishatskiy, S.; Wind, J.; Abetz, C.; Abetz, V. Pyrene-POSS nanohybrid as a dispersant for carbon nanotubes in solvents of various polarities: Its synthesis and application in the preparation of a composite membrane.*Nanoscale Res. Lett.* **2012**, *7*, 296.

26. Chen, P.; Wu, X.; Sun, X.; Lin, J.; Ji, W.; Tan, K.L. Electronic structure and optical limiting behavior of carbon nanotubes. *Phys. Rev. Lett.* **1999**, *82*, 2548–2551.

27. Kuo, A.C.M. Poly(dimethylsiloxane). In *Polymer Data Handbook*; Mark, J.E., Ed.; Oxford University Press: Oxford, UK, 1999.

28. Camino, G.; Lomakin, S.M.; Lazzari, M. Polydimethylsiloxane thermal degradation Part 1. Kinetic aspects. *Polymer* **2001**, *42*, 2395.

29. Chenoweth, K.; Cheung, S.; van Duin, A.C.T.; Goddard III, W.A.; Kober, E.M. Simulations on the thermal decomposition of a poly(dimethylsiloxane) polymer using the reaxff reactive force field. *J. Am. Chem. Soc.* **2005**, *127*, 7192–7202.

30. Camino, G.; Lomakin, S.M.; Lageard, M. Thermal polydimethylsiloxane degradation. Part 2. The degradation mechanisms. *Polymer* **2002**, *43*, 2011.

31. Briones-Herrera, J.C.; Cuando-Espitia, N.; Sánchez-Arévalo, F.M.; Hernández-Cordero, J. Evaluation of mechanical behavior of soft tissue by means of random laser emission. *Rev. Sci. Instrum.* **2013**, *84*, 104301.

32. Loomis, J.; King, B.; Burkhead, T.; Xu, P.; Bessler, N.; Terentjev, E.; Panchapakesan, B. Graphene-nanoplatelet-based photomechanical actuators. *Nanotechnology* **2012**, *23*, 045501.

33. Loomis, J.; Fan, X.; Khosravi, F.; Xu, P.; Fletcher, M.; Cohn, R.W.; Panchapakesan, B. Graphene/elastomer composite-based photothermal nanopositioners. *Sci. Rep.* **2013**, *3*, 1900.

34. Whitesides, J.M. The origins and the future of microfluidics. *Nature* **2006**, *442*, 368–373.

35. Whitesides, J.M.; Stroock, A.D. Flexible methods for microfluidics. *Phys. Today* **2001**, *54*, 42–48.

36. Szczukiewicz, S.; Magnini, M.; Thome, J.R. Proposed models, ongoing experiments, and latest numerical simulations of microchannel two-phase flow boiling. *Int. J. Multiph. Flow* **2014**, *59*, 84–101.

37. Colchester, R.J.; Zhang, E.Z.; Mosse, C.A.; Beard, P.C.; Papakonstantinou, I.; Desjardins, A.E. Broadband miniature optical ultrasound probe for high resolution vascular tissue imaging. *Biomed. Opt. Express.* **2015**, *6*, 1502–1511.

38. Colchester, R.J.; Mosse, C.A.; Bhachu, D.S.; Bear, J.C.; Carmalt, C.J.; Parkin, I.P.; Treeby, B.E.; Papakonstantinou, I.; Desjardins, A.E. Laser-generated ultrasound with optical fibres using functionalized carbon nanotube composite coatings. *Appl. Phys. Lett.* **2014**, *104*, 173502.

39. Velazquez-Benitez, A.M.; Hernández-Cordero, J. Optically controlled wavelength tunable fused fiber coupler. In Proceeding of the

Workshop on Specialty Optical Fibers and their Applications 2013, Sigtuna, Sweden, 28–30 August 2013.

40. Hill, K.O.; Meltz, G. Fiber Bragg grating technology fundamentals and overview. *J. Lightwave Technol.* **1997**, *15*, 1263–1276.

41. Cabriales, L.; Hautefeuille, M.; Fernández, G.; Velázquez, V.; Grether, M.; López-Moreno, E. Rapid fabrication of on-demand high-resolution optical masks with a CD–DVD pickup unit. *Appl. Opt.* **2014**, *53*, 1802–1807.

INDEX

A

ABC triblock copolymer, 231, 232, 233, 234, 235, 236, 237, 238, 241, 242, 243, 244, 245, 246, 247, 248, 249, 251, 254, 257
aniline-capped gold nanoparticles, 56

C

cellulose, 121, 122, 129, 131, 132, 133, 134, 138, 140, 141, 142, 143
Composites, 1, 118, 228, 262, 270, 274
conformation, 72, 77

D

DNA molecules, 55, 56, 58, 61, 62, 63, 64, 67, 69

E

Electrical Properties, 99, 102, 119
Electrochromic Property, 86
electrospray, 71, 72, 73, 74, 75, 76, 77, 78, 81, 83
emulsions, 4, 121, 122, 123, 127, 137, 138, 141

F

film feature, 32
Fracture, 1, 8, 9, 10

G

gold nanowires, 56, 58

H

hydrophilicity, 122, 134

M

Magnetoelectric nanocomposite, 170
Magnetostrictive, 170
mixed solvents, 32, 33, 48, 51
Morphology, 51, 119, 232, 253, 255, 257

N

NH_2-functionalized polymer thin films, 56, 58

O

oil–water separation, 122, 123, 134, 136, 138, 139, 140
Optical Properties, 102, 116
organic light-emitting diodes, 71, 72, 83, 86

P

PDMS-nanocarbon composites, 260
PECVD, 57, 58, 59, 61, 63, 64, 69, 158, 159, 160, 162, 164, 165
Phase diagram, 232, 241, 242, 244, 246
photomechanical actuation, 260, 279, 286
photothermal conversion, 260
photovoltaic, 31, 32, 33, 43, 45, 46, 48, 49, 50, 51, 52, 53, 73
Polyaniline, 67, 85, 86, 98, 99
polymer, 2, 3, 6, 7, 9, 11, 12, 17, 261, 267, 271, 274, 277, 278, 284, 288, 289, 290
Polymer Blends, 101, 102, 117
Polymer brush, 232

R

random copolymer, 31, 32, 33, 47, 258

S

Self-consistent field theory, 232
Surfaces, 1

T

Thermal properties, 1
Thin films, 1, 104

U

underwater superoleophobicity, 121, 122, 134
UV-Vis Absorption, 86

V

Vapor Deposition, 102